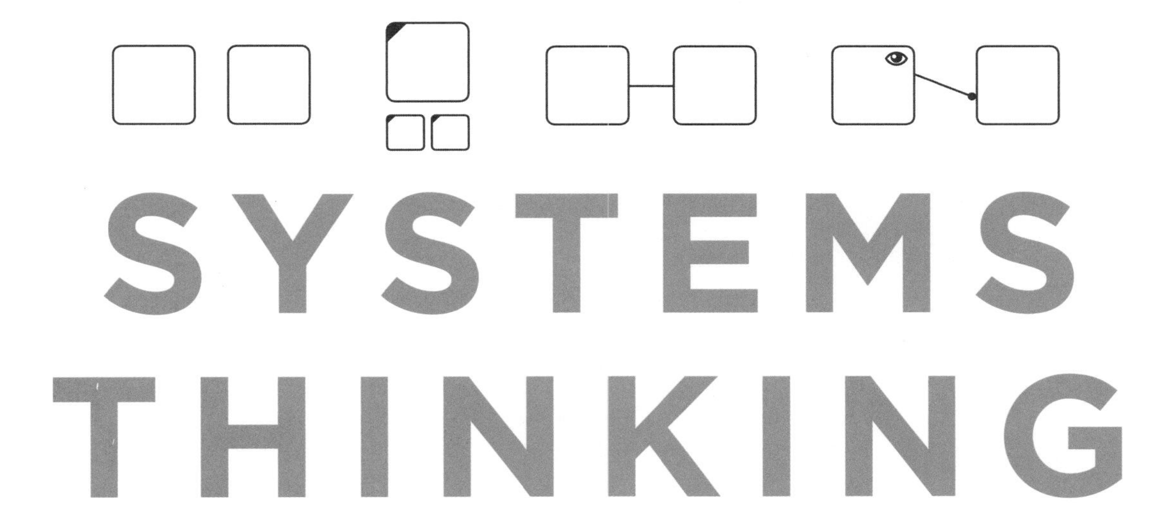

SYSTEMS THINKING

MADE SIMPLE

New Hope *for* Solving Wicked Problems

DEREK CABRERA & LAURA CABRERA

ODYSSEAN PRESS

All figures courtesy of authors unless otherwise credited in the list of Figures

Cover and book design by Michael Di Biase
Edited by Erin Powers

Printed in the United States of America
For information about or permission to reproduce selections from the book, contact: Cabrera Research Lab, Ithaca, NY
info@crlab.us
www.crlab.us

For all videos mentioned in the book, go to: crlab.us/stms

The ideas in this book can be applied using systems thinking and modeling software, go to: metamap.me

ISBN: 978-0-9963493-0-7

THANK YOU

We want to thank the many friends and colleagues who have helped us consistently improve our work, and partnered with us to transform systems and people into better thinkers. First, this book would not have been possible without the tireless efforts of our editor Erin Powers, who transformed our ramblings into sentences that will roll through your mind like thunder, and our designer Michael Di Biase, who used his considerable design aesthetic not merely to make a beautiful book but more importantly to increase deep understanding of concepts. We would also like to thank the various agencies that have funded our research over the years, including: USDA-NIFA, National Science Foundation, National Institutes of Health, Centers for Disease Control, and the Pennsylvania Department of Education. In particular the USDA-NIFA sponsored ThinkWater team has been highly engaged in this work, by providing insight and expertise into deepening the public's understanding of these concepts over the years. In particular, Jennifer Kushner and Deborah Hoard have been champions of change and invaluable partners in this effort. The rest of our remarkable team includes Kate Reilly, Peter Mason, Michael Di Biase, Erin Powers, Christopher Froehlich, Nicholas Campbell, Michelle Bloodworth, Catherine Bornhorst, and Carlos Romero. We have also benefited from the clear guidance of Jim Dobrowolski, Art Gold, Doug Parker, and Reagan Waskom. We'd like to thank our colleagues, Tom O'Toole at Cornell's Institute for Public Affairs, and Dennis Charsky at Ithaca College's Roy H. Park School of Communications. Most importantly, a special thank you to our students at both Cornell University and Ithaca College, who always push us to be better teachers and learners. Our friends and colleagues in the UK at the University of Hull, Management Systems, Gerald Midgley and Jennifer Wilby. There are also too many folks to count whom we have worked with over the years but a few that must be mentioned for their hard work, and dedication to helping this work impact their part of the world. Our friends in Fairfax County Public Schools: Maura Burke, Mary Ann Ryan, Jesse Kraft, Edith Dobbins, Debby Fulcher, Meghan Callahan, Ann Erickson, Christina Dickens, Carol Horn, and all the amazing teachers who are fully engaged in getting their kids to become systems thinkers. Our friends in Ithaca City School District: Luvelle Brown, Lee Ginenthal, Jason Trumble, Crystal Sessoms, Caitlin Redfield, and the amazing group of innovative ICSD teachers, Caren Arnold, Rebecca Baum, David Buchner,

Patricia Caughey, Mike Cecere, Tara Ciotoli, Kristin Devita, Courtney DeVoe, Duane Diviney, Valerie Evans, Kristin Herman, Louis Hicks, Jane Koestler, Michelle Kornreich, Mary-Elizabeth McDaniel, Kate Praisner, Eric Pritz, Ellie Rosenberg, Shane Taylor, Tracy Tidd, Emily Ufford, Bob Walters, and Kathleen White. Our friends at BLaST IU17: Cori Cotner, Bill Martens, Jerry Christy, Mark Burke, Amy Martell, Dan Martell, Christina Steinbacher-Reed, Heather McPherson, Micah Russell, Rosanna Hausammann, Amber Hildebrand, Rae Pitchford, Denise Cuevas, Rebecca Folk, Kristi Vergason Alexander, Cindy Gilson Young, Tracy Kahn, and George Ness, and all the other amazing educators at IU17 who are working towards their vision of 250,000 Thinkers. Our fantastic ThinkSTEM community including Heather Manchester, Kayla Eberth, Tim Miller, Kurt Risch, Brian Manchester, Jessica Catania, and all our rockstar ThinkSTEM students, who will be the future scientists we need. Our friends at Green Township School District, New Jersey: John Nittolo, Tara Rossi, Louis Rossi, Brian Martin, Karen Bessin, Aimee Castellana, Tara Lavalley, Jessica Wilds, and Mary Fox. Ted Caldwell for his tireless work on the MetaMap Platform. A few of the many smart and passionate innovators in the Silicon Valley we've had the privilege to work with, Tien Tzuo, Chris Cabrera, Evan Ellis, Steve DeMarco, Bernie Kassar, Scott Broomfield, Albert Lee, and Mike Lee. Scott Campbell at CAEL. And the many, many, others who have joined us on this journey for the last decade. We also must thank our academic mentors and advisors who have influenced us in so many ways without knowing it: Will Provine, Rosemary Cafarella, Jim Blanchard, Jerry Ziegler, Jennifer Gerner, Rosemary Avery, John Guckenheimer, and Steve Strogatz. Finally, we wouldn't have been susceptible to mentoring in the first place if it were not for our remarkable parents, Eduardo and Ruth Cabrera and Dottie Crago-Faust, whose lessons continue to teach us everyday and have been passed down to our own children, Elena, Gianna, and Carter. Thank you.

For Carter, Gianna, and Elena,
and to all our students who have helped us to translate systems thinking
into something simple without losing the fidelity.

SYSTEM OF CONTENTS

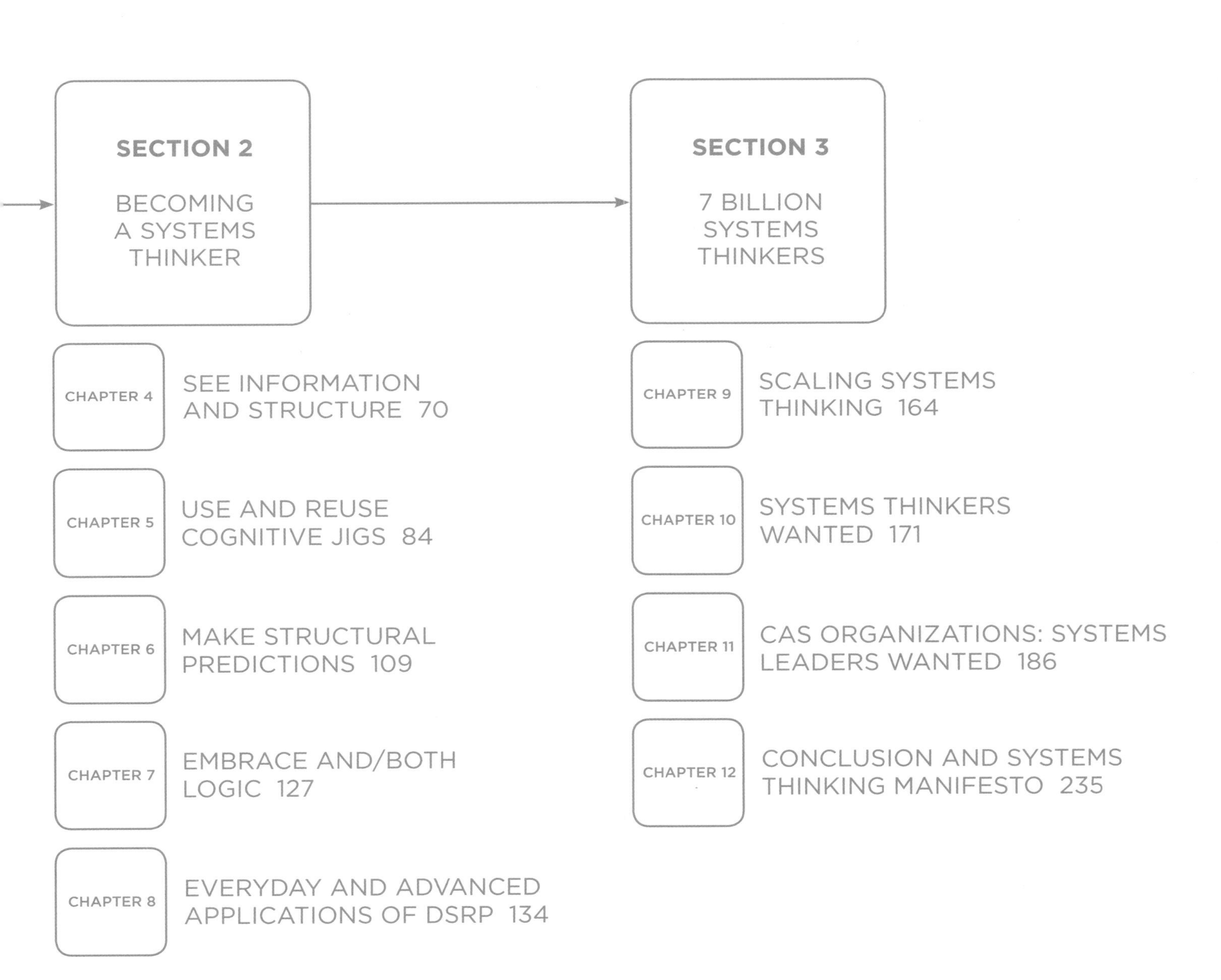
SECTION 2
BECOMING A SYSTEMS THINKER
SECTION 3
7 BILLION SYSTEMS THINKERS

SECTION 1

SIMPLE RULES OF SYSTEMS THINKING

SECTION 1
SIMPLE RULES OF SYSTEMS THINKING

Systems Thinking Made Simple doesn't mean that we're going to oversimplify it like a *... for Dummies* book. It means that we will show how systems thinking emerges when we focus on a simple set of rules. After years of searching for unifying principles, many experts and practitioners in the field of systems thinking have embraced DSRP as universal to all systems thinking methods.

Originally a complex mathematical formulation, DSRP has since been made more accessible through powerful modeling and visualization tools. There are two surprising things about our new understanding of systems thinking. First is how simple the four rules are:

DISTINCTIONS RULE: Any idea or thing can be distinguished from the other ideas or things it is with

SYSTEMS RULE: Any idea or thing can be split into parts or lumped into a whole

RELATIONSHIP RULE: Any idea or thing can relate to other things or ideas

PERSPECTIVES RULE: Any thing or idea can be the point or the view of a perspective

Perhaps more so than the simplicity of the rules, what is surprising is how these four simple rules can be mixed and matched, combined and recombined in ways that are immensely complex, leading to robust, systemic thinking. This unexpected connection between complex thought processes and simple rules has broad implications. We are astonished to learn that the breathtaking diversity and creativity of nature that produces peacocks, giraffes, and star-nosed moles is born of genetic mutations of the four nucleotides of DNA (ATCG). Much like the genetic code that underlies all species, DSRP provides a cognitive code that underlies human thinking. Systems thinking can increase our personal effectiveness as human beings, help us solve everyday and wicked problems, and transform our organizations. The discovery of DSRP means that we don't need to spend a lifetime becoming systems thinkers. We can begin by following the four simple rules out of which deep understanding and new insights and solutions will emerge. Simple is not the same as simplified or dumbed-down; smart arises from simple. Simple is sublime.

To truly effect change, we must democratize systems thinking, empowering everyone—not just the enlightened or privileged. In the chapters that follow, you'll learn why systems thinking is so important in every sector of society and to every individual. You'll also learn what systems thinking is and how to apply its underlying four simple rules. This is a book about the power of simple rules. In other words, small things done by many can lead to big changes.

CHAPTER 1

NEW HOPE FOR WICKED PROBLEMS

WHY SYSTEMS THINKING?

At its core, systems thinking attempts to better align how we think with how the real world works. The real world works in systems—complex networks of many interacting variables. Often nonlinear, complex, and unpredictable, real-world systems seldom correspond with our desire for simplistic, hierarchical, and linear explanations. Systems thinking is the field of study that attempts to understand how to think better about real-world systems and the real-world problems we face.

WICKED PROBLEMS RESULT FROM THE MISMATCH BETWEEN HOW REAL-WORLD SYSTEMS WORK AND HOW WE THINK THEY WORK. SYSTEMS THINKING ATTEMPTS TO RESOLVE THIS MISMATCH.

WHAT IS THE ROOT CRISIS?

Research[1] that surveyed the faculty of Cornell University to identify how scientists from different disciplines thought about the most pressing crises facing humanity asked, "One significant crisis humanity currently faces is...?" The faculty generated 116 unique crises. We then asked each faculty member to rank (on a scale of 1-5) these 116 crises in terms of both *importance* and *solvability* and to sort the 116 crises into their own categories. We applied multidimensional scaling and cluster analysis to their results to create 7 categories: environment and resources, health and disease, education and technology, influence, social institutions, human nature/perspective, and economics and poverty.

"Climate change and its effect on ecosystems" was ranked #1 in importance, whereas the crisis ranked most solvable was "loss of civil liberties in the US under the guise of fighting terrorism." The only crisis ranked in the top ten for both importance and solvability was "shortage of potable and clean water." Interestingly, the most important and pressing problems on the list were also deemed the least solvable, while the most solvable were deemed the least important.

At the conclusion of the study, we found ourselves wondering: Is there a "meta-crisis" that lies at the root of these varied crises or wicked problems? Yes. The root crisis is our *thinking*. Einstein is often paraphrased as having said, "A new type of thinking is essential if mankind is to survive and move toward higher levels."[2]

---

[1]Cabrera, D., Mandel, J. T., Andras, J. P., & Nydam, M. L. (2008). What is the Crisis? Defining and Prioritizing the World's Most Pressing Problems. *Frontiers in Ecology and the Environment*, 6(9), 469-75. doi: 10.1890/070185.; Forsythe, P. (2008, Nov). Beyond the Frontier: A Global Identity Crisis [Audio Podcast]. Available at: http://goo.gl/IHgKNQ

[2] (1946, May 25). Atomic Education Urged by Einstein. *New York Times*. While at the time Einstein was indicating the need for new thinking brought about by developments in atomic physics, modern day interpretation extrapolates his sentiments to mean more broadly, "The significant problems we face cannot be solved at the same level of thinking we were at when we created them."

Table 1.1: Most Important Vs. Most Solvable Crises

The 10 Most Important Crises	The 10 Most Solvable Crises
1. Climate change and its effects on ecosystems (4.39, 2.63)*	1. Loss of civil liberties in the US under the guise of fighting terrorism (4.02, 3.75)
2. Corporations have too much influence in governing (4.24, 3.35)	2. Death of children due to preventable causes (3.38, 3.35)
3. Lack of long-term perspective in political, environmental, and social actions (4.23, 2.69)	3. Inequitable access to healthcare (3.86, 3.77)
4. Humans are unsustainably exploiting the environment (4.13, 2.79)	4. Women's reproductive health, education, control, and options are dictated by others (3.85, 3.34)
5. Maintaining the health of the planet (4.1, 2.67)	5. Lack of sufficient education in science, critical thinking, and environmental issues (3.8, 3.87)
6. Lack of global responsibility on the part of corporations, governments, and individuals (4.03, 2.97)	6. The epidemic of preventable illnesses in the Third World (3.44, 3.81)
7. Global poverty and its effects (3.98, 2.48)	7. Children's activities are too structured and do not engage them in community (3.78, 2.64)
8. Inequitable distribution of wealth among people (3.97, 2.32)	8. People and governments are paying less attention to basic science research (3.74, 3)
9. Unsuitable growth in energy use (3.96, 2.95)	9. Shortage of potable and clean water (3.59, 3.94)
10. Shortage of potable and clean water (3.94, 3.59)	10. Growing obesity that impacts health by increasing the risk of chronic diseases (3.59, 2.89)

*Note: Parentheses show average importance and average solvability of crisis on a scale of 1 to 5

Problems are not divorced from the way we think about them. So what is the thinking style that has led to these sundry wicked problems? It is a style of thinking that is out of balance, it is:

- Overly focused on the parts (reductionism) to the exclusion of the whole (holism);
- Excessively hierarchical to the exclusion of more complex, distributed networks;
- Over reliant on static categories rather than part-whole groupings that result from perspectives;
- Overly linear and causal at the expense of seeing nonlinear webs of causality;
- Biased toward seeing structural parts but overlooking dynamic relationships; and
- Based on bivalent (2-states) rather than multivalent (many states) logic.

Climate change, hunger, wealth distribution, and childhood obesity, while all legitimate crises, are not the *root* crisis. The root crisis is the way we think. That is the problem that underlies all the other problems. We must change our thinking from the binary, linear, and categorical kind to a new form of thinking called systems thinking to solve the issues facing humanity.

The field of research called "systems thinking" was born of a similar question (What is the root crisis?). Today, many systems scientists believe that systems thinking is the way to solve wicked problems. As research scientists and systems theorists, we have made both a passion and a career out of taking a machete to the tangled undergrowth of the field of systems thinking. The term "systems thinking" suggests a crucial relationship between systems (the basic unit of how the natural world works) and thinking (the way we construct mental models of this world). Therefore, systems thinking must fundamentally balance what we know about real-world systems and what we know about the knower.

THE POPULARITY AND THE PROMISE

Systems thinking has gained popularity outside of academia, as evidenced by book sales of works like Peter Senge's *The Fifth Discipline*, the first mainstream book that brought systems thinking ideas to the general public. Touted as a means to gain personal mastery and solve organizational problems, Senge's 1990 book sold more than a million copies worldwide.

Today, the popularity and the promise of systems thinking hinges on the same desire to gain personal mastery to solve both everyday and wicked problems in our lives and in our organizations. Everyday problems are what they sound like, problems experienced everyday:

- By parents, teachers, students, employees, bosses, policy makers, scientists, and citizens;
- At any age: toddlers, teenagers, and adults;
- At work, home, school, or play.

Not thinking systemically about or ignoring everyday problems often makes them turn into wicked problems. Gerald Midgley, the preeminent historian in the field of systems thinking, explains what a wicked problem[3] is:

Wicked Problems involve...

- Many interlinked issues, cutting across the usual silos (e.g., economy, health and environment), making for a high degree of complexity;
- Multiple agencies (across the public, private and voluntary sectors) trying to account for multiple scales (local, regional, national and global);
- Many different views on the problem and potential solutions;
- Conflict over desired outcomes or the means to achieve them, and power relations making change difficult; and
- Uncertainty about the possible effects of action.

[3]Midgley, G. (2015) An Introduction to Systems Thinking. Integration and Implementation Sciences (I2S) http://goo.gl/ZT7emg Also see, Systems Thinking Conference at Cornell University crlab.us/stms/

What causes wicked problems? In a word, complexity. You have to take into account more stuff. There are more interrelations, more people involved, more disagreement, and less reliable information. You know the drill because these are the kinds of problems you deal with or think about everyday. Systems thinking has so much promise and is so popular because it offers us hope that we may be able to solve a few of these wicked problems.

These wicked problems aren't just for policy makers and presidents, either. How do you raise teenagers in a world that is fast and loose without stunting their growth or exposing them to unnecessary peril? How do you maintain your ethics and integrity in a world where lots of people get a leg up by cheating and manipulating? How do you remain calm, cool, and collected when that person makes you want to strangle them? How do you design a better widget or webpage when there are so many different types of users and uses? How do you know which strategy to take when you have so little reliable information or insufficient influence? These, too, are systems thinking problems: wicked problems.

If you're a scientist, academic, or just someone interested in the pursuit of cutting-edge knowledge, an equally important promise of systems thinking is that it offers a common language across methods, disciplines, and contexts, facilitating interdisciplinarity. The common desire to work together to understand and achieve great things is at the heart of scientific pursuit. However, those of us who create new understanding of the world through the pursuit of science often do so in different languages, using different specialized terminology and living in silos to such a degree that, if and when we ever do get together to work across boundaries in an interdisciplinary way, we have a hard time understanding each other.

Systems thinking gives us hope that the collective efforts of many to understand their little part of the world can come together to better understand the world as a whole. It gives us hope that we are going to be able to solve our most wicked problems.

DEMOCRATIZING SYSTEMS THINKING

Einstein said, "The whole of science is nothing more than a refinement of everyday thinking."[4] Much of our early work in systems thinking was funded by government institutions such as the National Science Foundation, National Institutes of Health, and United States Department of Agriculture to answer the question, how do we make a better scientist? No matter how many grants we received and how much progress we made, something gnawed at us. While systems thinking has the potential to advance the whole of science, it also has the power to transform everyone in their everyday thinking.

[4] Einstein, A. (1950). *Out of My Later Years: The Scientist, Philosopher, and Man Portrayed Through His Own Words* (p. 59). New York, New York: Philosophical Library.

To save our planet, solve crises, understand complex systems and their wicked problems, we don't just need better scientists who think more systemically, we need better *citizens* who think systemically.

Our research lab has been lucky enough to be one of the leading labs in the country in systems thinking research. Our lab focuses on three areas:

1. Advancing research on both the theory and practice of systems thinking;
2. Developing innovative media, tools, and technology for systems thinking; and
3. Facilitating public understanding of systems thinking.

This book is focused on all three areas: advancing public understanding of the research, tools, and technology of systems thinking.

Today, we are convinced that anyone in the world at nearly any age can understand and benefit from systems thinking and that their benefit would in turn help society. But this wasn't always the case. When Derek began, the basic theory was expressed in a way that was inaccessible and abstract:

$$ST_n = \bigoplus_{info} \bigotimes_{j \leq n} \{: D_o^i \circ S_w^p \circ R_r^a \circ P_v^\rho :\}_j$$

His research and equations[5] needed to be translated in order to make them more accessible to a wider audience. But he had no idea how to get from that string of symbols and complex ideas to something clear to someone regardless of their training, background, or age. Derek met Laura in the process of searching for an answer. She was an expert in a thing he had never heard of called "translational research." Translational research is the study of how to translate into practice abstract theories that can be used to better the real world. Over several decades we have spent our careers as researchers in the theory and practice of systems thinking. But for the last decade we worked in the trenches helping to build public understanding of systems thinking. This book is the translation of that string of symbols above: systems thinking made simple.

The thing we are most proud of is not the success of the theory Derek developed and its growing importance in the systems field. What's most inspiring is that today, preschoolers around the world are learning the same four simple rules of systems thinking that we teach doctoral students at Cornell University. That always puts a smile on our faces. It has been truly inspiring to see systems thinking "made simple" have such a profound impact on the lives of so many, from grade school to graduate school, in all walks of life and at every level of achievement.

[5]The equation explains that autonomous agents (information, ideas or things) following simple rules (D,S,R,P) with their elemental pairs (*i–o, p–w, a–r, ρ–v*) in nonlinear order (:) and with various co-implications of the rules (○), the collective dynamics of which over a time series *j* to *n* leads to the emergence of what we might refer to as systems thinking (ST).

WE CAN TEACH AND LEARN SYSTEMS THINKING

What this means is that not only can we answer with clarity the question, "what is systems thinking?," we can also teach it and all people can learn it.

We know this because the same visual mapping process that our K-12 students are using to better understand any subject matter, and more importantly, their own thinking, is also used to teach researchers, leaders, policy makers, academics, and business people all over the country.

While systems thinking started as a way to change how scientists thought about science, its future must have democratic aspirations. Systems thinking belongs to the people; it needs to be democratized by increasing public understanding of it. Because for systems thinking to truly work, all people need to better approximate reality, not just scientists with their controlled experiments, but *all citizens with their daily experiments*. Systems thinking can and will advance the most cutting edge discoveries in science, but it also helps everyone raise better kids, lead an organization, manage a better team, learn algebra, develop an ethical internal compass, and solve everyday problems. The goal of systems thinking should be nothing short of saving the world, one systems thinker at a time. This requires education. We do not mean to limit the definition of "education" to its traditional forms but rather extend it to all forms, from K-12 to adult, from formal to informal, from traditional to alternative, and from organism to organization. Human learning is what we are all doing (or not doing) everyday as we interact with the world.

That's why we wrote this book, as a form of education for anyone who wants to learn systems thinking. We wanted to show that systems thinking can be a powerful tool that is also accessible. Systems thinking can be made simple. The rest of the book provides:

1. An understanding of the field of systems thinking;
2. The changes we see in what we call systems thinking version 2.0;
3. The application of complexity science to our understanding of systems thinking;
4. The four simple rules that are universal to all the different methods of systems thinking;
5. The new logic underlying these rules;
6. Step-by-step instruction in visualizing and modeling systems thinking; and
7. Specific uses and models to demonstrate its application;
8. How to scale systems thinking in individual people and organizations;
9. How systems thinkers develop an internal compass; and
10. How systems leaders can lead and manage systems thinkers within organizations.

Over decades of studying systems thinking, it has become clear that it has enormous potential to change society and to tackle some of our most wicked problems and to help us grasp some of the most complex systems in the universe. *But systems thinking without systems thinkers will change nothing.* We believe that if you can't explain something simply, you don't understand it well enough. We must push ourselves to make simple clear sense. A better world will come when we educate, in the broadest sense of the term, 7 billion systems thinkers. We hope that you'll be one of them.

SYSTEMS THINKING WITHOUT SYSTEMS THINKERS WILL CHANGE NOTHING. FOR SYSTEMS THINKING TO TRULY WORK, ALL PEOPLE NEED TO BETTER APPROXIMATE REALITY NOT JUST SCIENTISTS WITH THEIR CONTROLLED EXPERIMENTS, BUT CITIZENS WITH THEIR DAILY EXPERIMENTS.

CHAPTER 2
THE SIMPLICITY THAT DRIVES COMPLEXITY

SYSTEMS THINKING V1.0

Before we address the exciting promise of systems thinking v2.0, or systems thinking made simple, we must briefly summarize its predecessor, version 1.0.

What is systems thinking? As a doctoral student, Derek asked the same question and was shocked by just how many different and conflicting answers there were. He asked a simple question but the experts couldn't seem to provide the kind of simple explanation that so often indicates deep understanding. Dogged in his pursuit, Derek decided to take a look at all the possible answers to this question and see if he could make better sense of them all.

First, he needed to establish some kind of boundary for systems thinking. In other words, he needed to figure out what was *inside* the universe of answers and what was *outside*. After extensive review, Derek coined the term *MFS Universe of Systems Thinking*[1]—an acronym for Midgley, Francois, and Schwartz—three systems thinking experts who made significant contributions to the field by doing some of their own synthesis of relevant works in systems thinking. Consequently, the MFS universe represents a thorough literature review of the field.

[1]Midgley, G. (2003). *Systems Thinking*. Thousand Oaks, CA: SAGE Publications.; Francois, C. (2004). *International Encyclopedia Of Systems And Cybernetics* (2nd ed.). Munich: K G Saur.; Schwarz, E. (1996). *Streams of Systemic Thought*. May 2001 ed. Neuchâtel, Switzerland. Retreived from: http://goo.gl/Jdmf-GK; Cabrera, D. (2006). *Systems Thinking*. Cornell University, Ithaca, New York.

Gerald Midgley, a systems thinking expert and notable historian on the subject, edited a four-volume set of 97 different methods of and approaches to systems thinking. Charles Francois compiled an encyclopedia with 3,800 entries of systems thinking concepts, theories, methods, and frameworks. Eric Schwarz created a network diagram with 648 nodes (and thousands of connections), each representing a famous systems thinker, theory, or method of systems thinking.

Figure 2.1: MFS Universe of Systems Thinking

Together, these three sources provide a relatively comprehensive review of all the possible answers to the question, what is systems thinking?, and represent lifetimes of study. What kinds of things are in the MFS Universe? The problem is that it's a kind of big tent pluralism where any guest is welcome. Here's just of a few of the types of things you might find:

- Formal theories such as network theory, chaos theory, or general systems theory;
- Important concepts such as unintended consequences;
- Approaches designed for specific purposes such as soft systems methodology (group process); and
- Modeling methods for building models of systems such as system dynamics.

It really is a menagerie of varied guests. And for systems thinking theorists and those interested in the history of the field like Derek, it's important and fun to explore all of these types of systems thinking. We can look at specialized applications (what they're good at and not good at), what's hot and trending, what's remained static, what theories and practices have gone cold, and which ones are historically significant, even if now considered invalid.

But for practitioners, this is not only quite frustrating, but utterly impractical. Nevertheless, it does help practitioners to see the big picture and to see some of the emerging trends and big ideas in the field. Figure 2.2 shows the most popular theories, frameworks, and methods in systems thinking. The axis shows that systems thinking traverses many fields from the physical and life sciences to the social sciences and everything in between. A systems thinking method that is well known by a physicist, chemist, biologist, or engineer might be completely foreign to a cognitive scientist, sociologist, economist, or human ecologist, and vice versa. The most promising methods apply across disciplines.

It is also important to consider the real-world usage of the term systems thinking. A search for “systems thinking” on careerbuilder.com yields 73 jobs in 30 days that list systems thinking as a skill in the job description. Here's just a few:

- Senior Instructional Designer in Williamsville, NY
- Fuel Inventory Manager in West Des Moines, IA
- Entry Level Production Engineer in Wilmington, MA
- Inside Sales Engineer in Boulder, CO
- Senior Business Process Specialist in Milwaukee, WI
- College Vice President in Galveston, TX
- CEO for a tech company in San Francisco, CA

What does it mean in practice to be a systems thinker? How does one develop this skill? Neither the MFS Universe of systems thinking nor the ubiquitous use of the term in job ads provides us with an answer to these questions. Let's explore why. By way of analogy, imagine us considering the question a biologist might ask, what is life? It would be reasonable to gather up a relatively complete collection of all living species (birds, plants, mammals, fungi, bacteria,

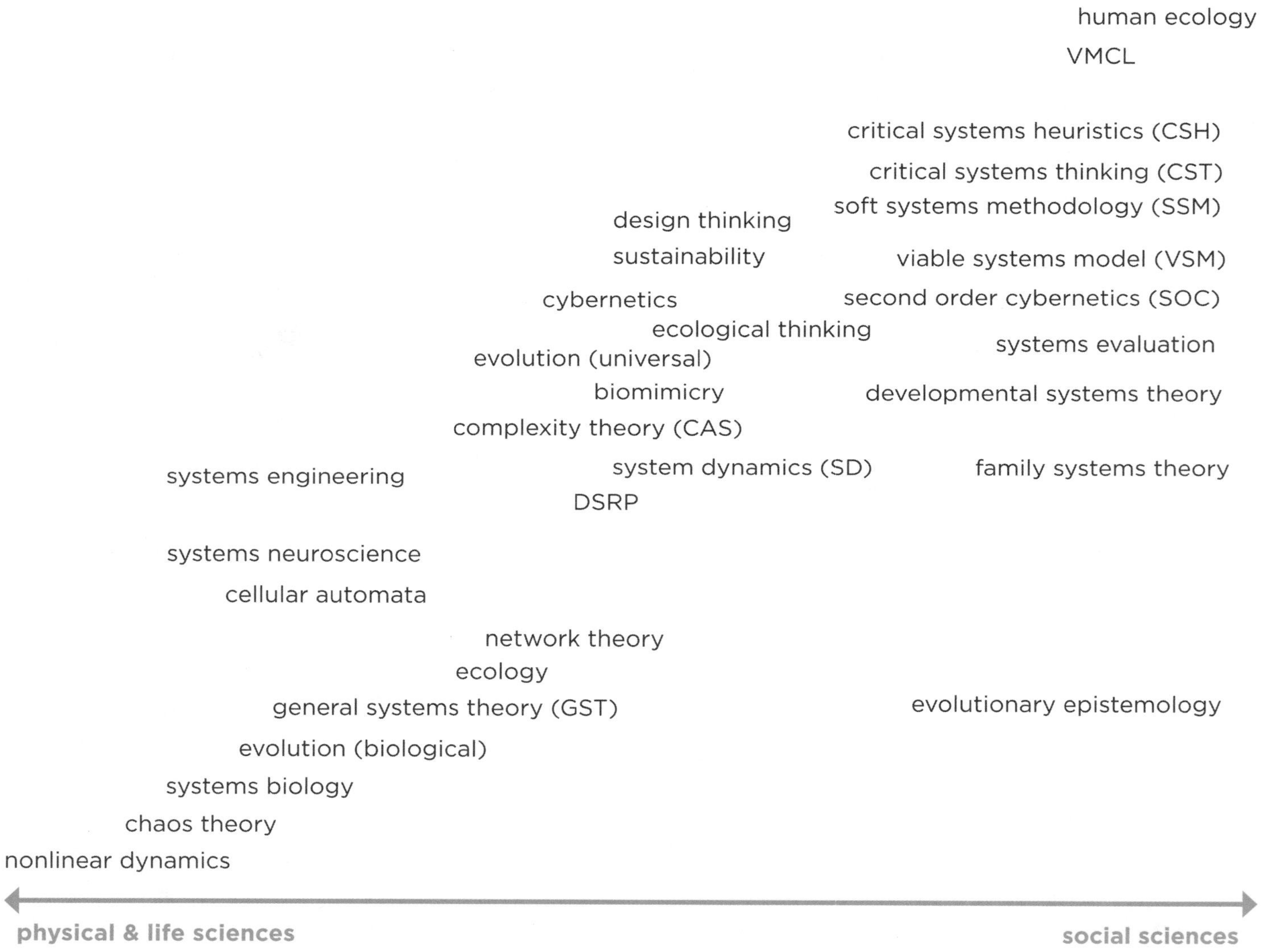

human ecology

VMCL

critical systems heuristics (CSH)

critical systems thinking (CST)

soft systems methodology (SSM)

design thinking

sustainability

viable systems model (VSM)

cybernetics

second order cybernetics (SOC)

ecological thinking

systems evaluation

evolution (universal)

biomimicry

developmental systems theory

complexity theory (CAS)

systems engineering

system dynamics (SD)

family systems theory

DSRP

systems neuroscience

cellular automata

network theory

ecology

general systems theory (GST)

evolutionary epistemology

evolution (biological)

systems biology

chaos theory

nonlinear dynamics

physical & life sciences

social sciences

Figure 2.2: Select Elements of MFS Universe of Systems Thinking

etc.) and establish them as a "universe of living things." Our specimen collecting is a good first step that establishes the boundary between living and nonliving things. Now the hard work begins. Looking at all these living things, we might try to find a pattern that cuts across all of them. We might say, they all have legs, but then immediately learn that some of them don't and therefore rule that out. We might say, they all metabolize, or they are all carbon-based, and so on. Over time, through a process of elimination, we would discover the necessary and sufficient conditions for life: an organism with organized structure and function capable of metabolism, growth, response to stimuli, adaptation, and reproduction.

The problem with the complicated morass of phenomena/elements in systems thinking is that big tent pluralism obscures what systems thinking *is*. It describes a bunch of *examples* of systems thinking.

Answering the question, "what is systems thinking?" with a litany of examples of systems thoughts, methods, methodologies, approaches, theories, and ideas, is like answering the biological question, "what is life?" with examples of plant and animal species. To answer the question we need to delve deeper and look for patterns that connect these examples of systems thinking.

For systems thinking to solve the wicked problems it was designed to solve, problem solvers need to know what it is. Moreover, if we identify its underlying patterns, systems thinking will be readily accessible to most people, so they can apply it to their lives, jobs, and problems.

Answering the question, "what is systems thinking?" with a litany of examples of systems thoughts, methods, methodologies, approaches, theories, and ideas, is like answering the biological question, "what is life?" with examples of plant and animal species. To answer the question we need to delve deeper and look for *patterns that connect* these examples of systems thinking.

Big tent pluralism was the state of the field as of ten years ago. This is systems thinking version 1.0, and it has enormous potential. However, this potential will never be realized if the problem solvers of the day can't grasp it. Each systems thinking v1.0 theory and method represents a specialty use or tool. But introducing people who want to solve problems to specialized tools that may not be ideally suited to their needs can lead to frustration and losing hope.

As a field, we have missed an opportunity to educate because we confuse tools for skills. Many people have given up on systems thinking because they didn't understand it *before* being introduced to a specialized tool. We can't tell you the number of people we've had to talk down from the ledge who said they invested a lot of time in system dynamics modeling (or some other specialized method) only to learn it wasn't the tool they needed for the job. To the untrained user, the failure of a method to address their problems equates to the failure of the field of systems thinking.

It also makes no sense to us that in order to understand systems thinking—the principles of which are deeply democratic—you have to receive elite training over decades. That seems aristocratic, not democratic, and a slap in the face to the philosophical foundations of the field. So we decided that we would understand systems thinking well enough to explain it simply to others. We relied on the fact that Derek had already proposed a theory that would come to be regarded as the universal principles that underlie the plurality of methods we see in the MFS universe.

SYSTEMS THINKING V2.0

There are two prerequisites to deep understanding of systems thinking v2.0. The first is the idea of mental models and the constant role they play in our everyday lives. The second prerequisite is to better understand the concept of a complex adaptive system (CAS), because for systems thinking to be successful, it must be adaptive. In this chapter, we will explore these two perquisite ideas and in chapter 3, we will explore the four simple rules that underlie all of the systems thinking methods and approaches we discussed that are part of the MFS universe of systems thinking (i.e, systems thinking v1.0).

Although it may seem obvious, it warrants stating that in the term *systems thinking*, *systems* is an adjective *describing* the noun, *thinking*. In other words, systems thinking is about thinking. Ironically, this fact has eluded many systems theorists and largely eluded the field for decades.

The shift toward systems thinking v2.0 addresses this. Much of systems thinking version 1.0 focused on the *systems* part of systems thinking because understanding and advocating for a *systems view* was the original focus of the field. The early theorists who did focus on thinking were hampered by the fact that *thinking* (cognition, etc.) as a field of study was in its infancy.

Systems thinking v2.0 is the result of new scientific discoveries in our understanding of both *thinking* and *systems*. As Figure 2.3 depicts, systems thinking v2.0 balances the *systems* and the *thinking* parts of systems thinking. Systems thinking v2.0 balances the cognitive, emotional, and motivational factors that cause our mental models to be so out of alignment with real-world systems (i.e., our mental models are wrong). It requires that we become equally adept at human psychology and understanding the mind as we are at the physics, chemistry, and biology of real-world systems. Systems thinking v2.0 takes into account not just what is perceived about the real world, but also the perceiver's predilection to misinterpret. Systems thinking v2.0 takes into account our flaws as thinkers.

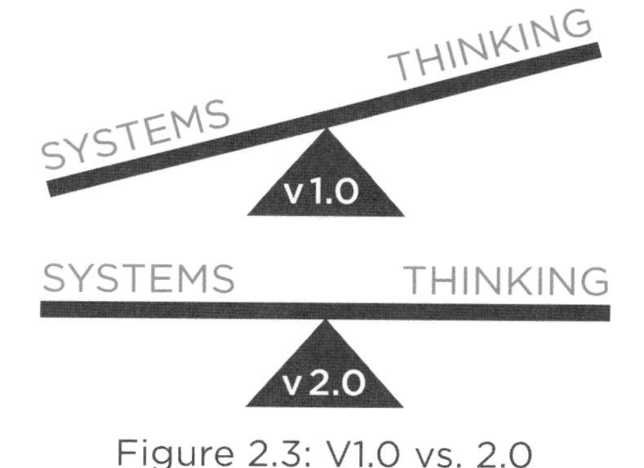

Figure 2.3: V1.0 vs. 2.0

Although it may seem obvious, it warrants stating that in the term *systems thinking*, *systems* is an adjective *describing* the noun, *thinking*. In other words, systems thinking is about thinking. Ironically, this fact has eluded many systems theorists and largely eluded the field for decades.

Table 2.1 contrasts the types of questions we tend to ask in systems thinking v1.0 versus v2.0.

TABLE 2.1: Systems Thinking Questions

V 1.0	V 2.0
What are systems?	What is systems thinking?
How do systems work?	How does systems thinking work?
Are there universal elements to systems behavior across different types of systems?	Are there universal elements to systems thinking regardless of approach?
What are the fundamental elements of a system?	What are the fundamental elements of systems thinking?
What are the simple rules of complex systems?	What are the simple rules of systems thinking?

Table 2.2 presents some of the fields that influence systems thinking.

TABLE 2.2: Influences on Systems Thinking

V 1.0	V 2.0
physics, chemistry, biology, mathematics, ecology, philosophy, engineering, organizational development	cognitive science, learning science, evolutionary epistemology, embodied cognition, metacognition, education, methodology, the sciences of meaning and understanding and methodology of pursuing knowledge, complexity, network theory

Systems thinking v2.0 enables us not only to understand the real world, but also the fascinating world inside our minds—the world of subjectivity, understanding, meaning making, thinking, the creation and evolution of knowledge, and learning itself.

SYSTEMS THINKING IS THINKING

We desperately need to get better at building mental models. Mental models aren't just fanciful ideas, they are often, like the mental models of the two wildebeests (in Figure 2.4, see video), a matter of life and death.

Figure 2.4: Mental Models

Watch Wildebeests' Mental Models: crlab.us/stms

Think of it this way. If we invented human flight but 6 out of 10 departing flights crashed before reaching their destination, we would have some work to do. Our mental models of flight—aerodynamics— were evidently flawed, based on our low success rate. The crashes are feedback that our mental models are out of alignment with reality because obviously, if we could prevent them, we would have.

Systems thinking v2.0 enables us not only to understand the real world, but also the fascinating world inside our minds—the world of subjectivity, understanding, meaning making, thinking, the creation and evolution of knowledge, and learning itself.

Whenever we don't get the results we want, whenever the behavior of a system surprises us, whenever the treatment doesn't solve the problem, it's the real world giving us feedback that there's something wrong with our mental models. The 2014 loss of two Asian airliners full of people is a tragic and wicked problem. The sum total of our existing mental models regarding the complex system surrounding these accidents is out of alignment with the real world. Although it seems simple that planes shouldn't just fall out of the air, the fact that they still do is a systems thinking problem.

The original intent of systems thinking was to get better at understanding the real world by coming up with new and improved mental models of the real world. To a much lesser extent, systems thinking also asked deep and penetrating questions about the mental models themselves. Systems thinking v2.0 is predicated on more closely aligning our current mental models with the real world.

Let's take a look at what we mean by mental models. We understand the real world through mental models. Here are three examples of mental models of real-world systems:

- Kepler (Earth-centered polyhedral orbits),
- Ptolemy (Earth-centered), and
- Copernicus (sun-centered).

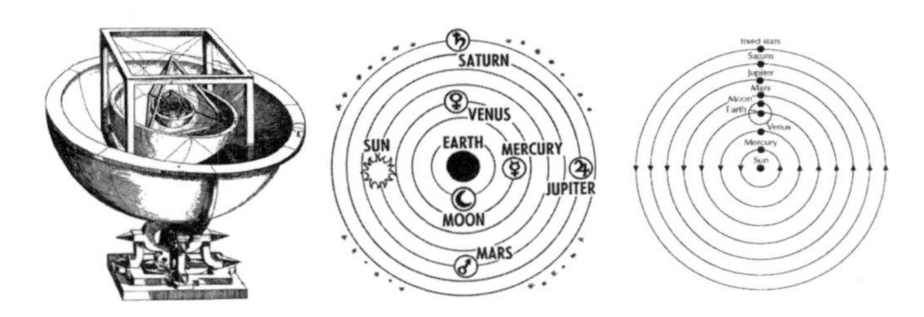

Figure 2.5: Mental Models Evolve: Kepler, Ptolemy, and Copernicus

TABLE 2.3: Some Economic Mental Models

	Classical	Neoclassical	Marxist	Developmentalist	Austrian	Schumpeterian	Keynesian	Institutionalist	Behaviouralist
The economy is made up of...	classes	individuals	classes	no strong view, but more focused on classes	individuals	no particular view	classes	individuals and institutions	individuals and organizations and institutions
Individuals are...	selfish and rational (but rationality is defined in class terms)	selfish and rational	selfish and rational, except for workers fighting for socialism	no strong view	selfish but layered (rational only because of an unquestioning acceptance of tradition)	no strong view, but emphasis on non-rational entrepreneurship	not very rational (driven by habits and animal spirits); ambiguous on selfishness	layered (instinct - habit - belief - reason)	only boundedly rational and layered
The world is...	certain ("iron laws")	certain with calculable risk	certain ("laws of motion")	uncertain, but no strong view	complex and uncertain	no strong view, but complex	uncertain	complex and uncertain	complex and uncertain
The most important domain of the economy is...	production	exchange and consumption	production	production	exchange	production	ambiguous, with a minority paying attention to production	no strong view, but puts more emphasis on production than do the Neoclassicals	no strong view, but some bias to production
Economies change through...	capital accumulation (investment)	individual choices	class struggle, capital accumulation and technological progress	developments in productive capabilities	individual choices, but rooted in tradition	technological innovation	ambiguous, depends on the economist	interaction between individuals and institutions	no strong view
Policy recommendations	free market	free market or interventionism, depending on the economist's view on market failures and government failures	socialist revolution and central planning	temporary government protection and intervention	free market	ambiguous - capitalism is doomed to atrophy anyway	active fiscal policy, income redistribution towards the poor	ambiguous, depends on the economist	no strong view, but can be quite accepting of government intervention

Source:www.zerohedge.com/sites/default/files/images/user3303/imageroot/2014/06/20140625_econ.jpg

Barbie represents another mental model. Barbie has become the dysfunctional mental model of many young girls (and boys) for how the female human body should look. Next to Barbie is a new crowd-funded doll called Lamilly who is based on the actual average dimensions of the female body. You might also notice that the clothing that comes with both dolls represents another mental model. Something as simple as a child's doll that reifies a mental model of girls can affect their notion of femininity and their self-esteem for a lifetime.

Figure 2.6: Lamilly and Barbie represent divergent mental models

Table 2.3 provides yet another example of mental models: nine schools of economic thought, each a large-scale mental model, that shape an entire academic discipline. The table shows the perspectives of each of the mental models (e.g., Classical, Neoclassical) on economic factors such as what the economy is made up of, whether the world is simple and certain or complex and uncertain, and how economies change. One interesting note about this table is that as one moves from left to right, the perceived complexity of the world that these mental models reflect seems to increase.

For example, the row labeled "The world is..." is first described as certain (with "iron laws") and then becomes uncertain and then later uncertain and complex. The table itself is one large mental model (economics), whereas each of the columns is another smaller-scale mental model, and each cell is an even smaller mental model causing one to wonder, what isn't a mental model? The answer is nothing. Everything we think is a mental model. Although it feels to our conscious self that we interact with the real world directly, in fact we interact *indirectly* with the real world through our mental models of it.

Mental models are all around us. They can be simple or wildly complex. They can describe important or unimportant phenomena. Mental models all try to explain or relay some meaning, presumably about the nature of our reality. And when we say reality it might be a tiny slice of reality like how to better cook a scrambled egg, or what is causing a guy at work to behave in such a mean-spirited way, or why racism persists, or why some people think it doesn't. Or reality might be how to better construct a democratic society, how to educate our children better, or how to think more systemically about anything and everything.

GETTING REALITY RIGHT

The British mathematician George E.P. Box said, "All [mental] models are wrong; the practical question is how wrong do they have to be to not be useful."[2] It's a simple statement with profound implications. It means that everything we think about the systems around us is merely an approximation. Our mental models are always wrong in the sense that they never completely capture the complexities of the real world. But mental models are useful because sometimes they get it "right enough." The Nobel laureate Herbert Simon coined the term *satisficing*[3] and contrasted it with optimizing. Simon explained that organisms of all kinds may avoid the costs associated with optimization (since perfection is expensive) and instead try to find a satisfactory solution. They're okay with good enough because it gets the job done.

In his book, *Thinking, Fast and Slow*, Nobel laureate and psychologist Daniel Kahneman describes his research on two modes of thought he calls System 1 and System 2. He writes, "System 1 operates automatically and quickly, with little or no effort and no sense of voluntary control. System 2 allocates attention to the effortful mental activities that demand it, including complex computations."[4] His purpose is to show how biased our human judgment can be. System 1 is evolutionarily valuable because although it's often wrong, it's right enough (i.e., cognitive satisficing) and has the added advantage of being fast. System 2 is more accurate but takes longer. System 1 works reasonably well when the problems humans deal with are routine, familiar, tangible, rudimentary, and simple (i.e., the kind of problems we might experience in hunter/gatherer, agrarian, or industrial ages). But as society becomes more complex and abstract, with more information, interconnections, perspectives, and systemic effects, the relative accuracy of System 1's mental models declines. Also, the amount of time we have to think about complex problems shrinks. What we need is System 2 accuracy with the speed of System 1. As the stakes get higher, it becomes even more important that our mental models reflect reality. The four simple rules of systems thinking that you will learn more about in chapter 3 provide a basis for better understanding System 1 and System 2 processes and increasing their accuracy and speed, respectively.

---

[2]Box, G. E. P. & Draper, N. R. (1987). *Empirical Model Building and Response Surfaces* (p. 424). New York, New York: John Wiley & Sons.; This same quote is often shortened to "all models are wrong, but some are useful" (p. 424).

[3]Simon, H. A. (1956). Rational Choice and the Structure of the Environment. *Psychological Review*, 63(2), 129-138. doi:10.1037/h0042769. (page 129: "Evidently, organisms adapt well enough to 'satisfice'; they do not, in general, 'optimize'."; page 136: "A 'satisficing' path, a path that will permit satisfaction at some specified level of all its needs.")

[4]Kahneman, D. (2011). *Thinking, Fast and Slow*. New York, New York: Farrar, Straus and Giroux. p. 20

Figure 2.7[5] shows the relationships among the variables we are dealing with as we attempt to solve real-world wicked problems. We create mental models that summarize and are capable of describing, predicting, and altering behavior. In other words, our mental models may lead us to think certain things about the real world and result in actual behaviors in the real world. These predictions, descriptions, or behaviors lead to real-world consequences that in turn provide feedback or data that inform our mental models. If we are paying attention, this feedback helps us adapt and select the best mental models. Ideally, we want our mental models to reflect the salient aspects of the real-world system or problem we are trying to solve. The way we know whether or not our mental model is right (or at least satisfices) is that we try it out in the real world and see what happens. If what we expect to occur occurs, then the feedback we receive from the real world tells us our mental model is well constructed. If we expect something to occur and it doesn't, then the feedback we receive from the real world tells us our model needs some work. Either way, we (hopefully) take in this new data which informs what is (hopefully) a continuous process of mental model improvement.

[5]Adapted from Gell-Mann, M. (1995). The Quark and the Jaguar: Adventures in the Simple and the Complex. New York, New York: W.H. Freeman and Co. Gell- Mann uses the term schema as a general term to refer to any model. Here we are using mental model to refer to conceptual models and schema to refer to general models of any kind including conceptual or mental models. Gell-Mann also includes data sources, which one presumes come from the real world and therefore can be simplified.

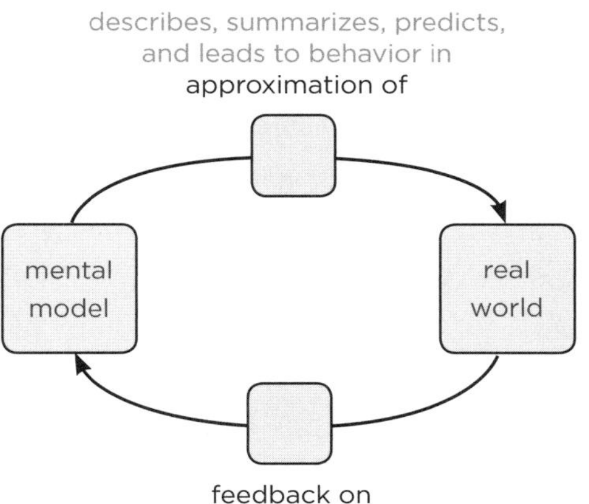

Figure 2.7: Mental Models Approximate the Real World, Which Provides Feedback to Adapt our Mental Models

But the basic process described in Figure 2.7 need not be thought of as purely a mental, cognitive, or conceptual process. It describes a process that has many synonyms that at first we might not perceive as such. Figure 2.7 also describes the processes of learning, evolution, feedback, adaptation, knowledge, science, and complex adaptive systems. For example:

- *Individual learning*: a process of building mental models and adapting them to the environment;
- *Organizational learning*: a process of sharing mental

models across individuals and adapting the mental models to the environment;

- *Feedback*: the process of acting upon and reacting to a stimulus, environment, or system;
- *Adaptation*: the process of changing to become better suited to one's environment;
- *Complex adaptive system (CAS)*: a system that adapts to become better suited to its environment;
- *Science*: a process of building mental models (knowledge) that better approximate reality; and
- *Knowledge*: the collection of individual or shared mental models used to navigate the real world.

If we simply exchange the term mental model with schema (a more general term that includes physical, chemical, biological, mental, or any other types of models) you can see that the model need not be a conceptual one, but could instead be a genetic model such as the DNA sequence of a platypus. Like mental models which represent a hypothesis, every organism and species is also hypothesis—a genetic model—that may or may not turn out to be viable, and therefore experience selective effects and compete with other models (a.k.a., evolution). A mutation is in one sense a model or hypothesis that may turn out to be adaptive or maladaptive. If ants live in deep tiny holes and adaptations lead to longer, pointier beaks in birds, then the model turns out to be adaptive, whereas the competing pancake-shaped beak model may turn out to be maladaptive.

Figure 2.7 describes learning as a global phenomenon as well as the basic structure of all evolutionary processes where things adapt to their environment. Figure 2.7 also describes learning at the societal level where science is an adaptive process of building mental models (knowledge) that better approximate reality.

These phenomena that may at first seem very different are structurally similar underneath. Each of these phenomena fundamentally involves a model or schema that gets tested against the real world. Selection pressures affect the viability of any model against competing models. There are winners and losers. Science works the same way. We come up with hypotheses, concepts, models, or theories and test them against the real world and see which ones win and lose. Learning, both individual and organizational, entails the development of mental models; the ones that work (for some purpose which is often not conscious to us at the time) survive. Those that don't work perish. Of course, this can be quite complex as you might develop mental models with the purpose of keeping yourself in denial, because the truth is simply too difficult to handle. You test the model in reality, and if it works, the model is preserved. So the purpose of the model is often important and the purpose is not always about getting closer to the truth. Figure 2.7 depicts how complex adaptive systems work, and describes the learning process that makes a complex adaptive system *adaptive*. The process that allows it to adapt to its environment is that it *learns*. Figure 2.7 just as easily characterizes all forms of adaptive knowing and learning. That is, it is the model of science itself. Remember, systems thinking is all about increasing the probability of getting the mental model right! Here "right" means that it better approximates the real world.

Individual survival is based on this feedback loop between mental models and the real world. But whole organizations and civilizations rely on it, too. Every organization's survival depends on its ability to learn. CEO Jack Welch said it this way: "An organization's ability to learn, and translate that learning into action rapidly, is the ultimate competitive advantage." So this model is critical. We want you to understand it so well that its basic structure is burned into your mind because it is the basis of systems thinking: that the mismatch between how real-world systems work and how we think they work leads to wicked problems. We want you to take another look at Figure 2.7, understand it, and then shrink it down into a little representational picture like this:

We can even shrink it down further into what we call a *sparkmaps*[6], which is a tiny, structural mental model that sparks prior understanding and can be situated in text like this ( ).

[6]Tufte, E. (2006). *Beautiful Evidence*. Cheshire, Connecticut: Graphics Press. Yale professor Edward Tufte developed "sparklines" to describe "intense, simple, word-sized graphics." Sparkmaps are intense, simple, word-sized metamaps that show the meta-cognitive structure of an idea and are used to reference concepts in the text and to aid memory and rebuilding of concepts.
Here's an example of a sparkline: respiration 9

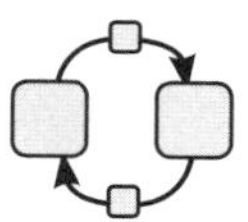

Figure 2.8: Sparkmap of Mental Model

Murray Gell-Mann is one of the greatest minds you've probably never heard of. His scientific achievements include decades of innovation; he gave us insight into the quantum world, for which he won a Nobel Prize. Gell-Mann, along with other great minds, founded the Santa Fe Institute for the study of complex systems (SFI), which has significantly advanced the field of complexity science. Figure 2.9 is a diagram he developed that gives a big-picture view of what he calls "Some complex adaptive systems on Earth."[7]

We'll start at the bottom, in the primordial soup, where pre-lifelike chemical reactions are subject to evolutionary pressures. Out of these processes emerges the biological evolution of both organisms and ecosystems. From this in turn emerges both the mammalian immune system and individual learning and thinking. This is an important step because this is where systems thinking lives. So pay particular attention to the red parts of the diagram [emphasis ours]. It is through the transmission of learning among individuals (both within and across generations) that human culture emerges. This

7 Figures 2.9 and 2.10 were adapted from Gell-Mann, M. (1995). *The Quark and the Jaguar: Adventures in the Simple and the Complex* (p. 20). New York, New York: W.H. Freeman and Co.

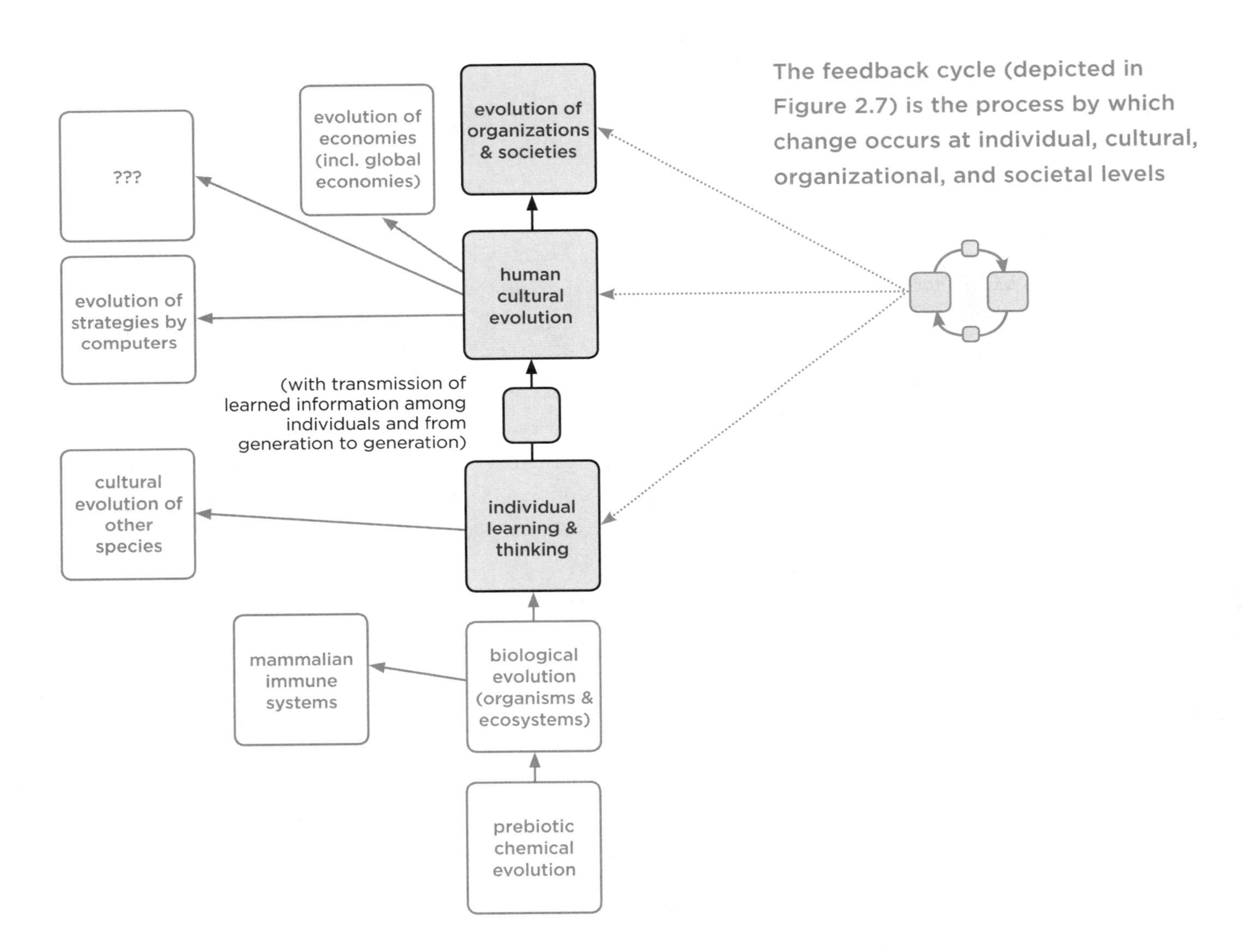

Figure 2.9: All Complex Adaptive Systems Rely on Feedback Between Schema and Reality

is really critical, because individual thinking and learning ( ) is at the *root* of the creation of culture (i.e., sharing mental models with others). At the cultural and organizational levels, the same learning feedback loop occurs but for the larger group or organization ( ). Of course, this same culture-building process leads to the emergence of economies, technologies, and other phenomena on a mass scale.

SYSTEMS THINKING IS A COMPLEX ADAPTIVE SYSTEM (CAS)

Let's revisit the idea that systems thinking, at its core, posits that the problems we face—the wicked problems we'd most like to resolve—are most often attributable to the mismatch between the way real-world systems work and the way we think they work. For example, a wicked problem might be world hunger, yet there is enough food to feed the world (2640 kcal/person/day of per capita food availability),[8] so this issue is resolvable. The solution requires a deeper understanding of the varied systems that contribute to world hunger, and that can be difficult, but it is not the case that we need to transgress the laws of physics in order to solve the problem of world hunger. The reason we can't solve world hunger is because we don't have a mental model that accounts for the social, economic, political, motivational, and cultural issues that shape the problem. Someday we will develop this mental model and world hunger will no longer be a wicked problem.

In other words, systems thinking is about building mental models that better align with real-world systems than those created under a non-systems thinking approach. The process of systems thinking results in a product: mental models. These mental models are representations, approximations, guesses, hypotheses, biases, or predictions about the real world. So systems thinking must be a form of thinking that is somehow qualitatively better at helping us build our mental models. Otherwise, its just plain old thinking, with all its sundry biases and issues.

Wicked problems are deemed so for a reason: the systems that lead to them are complex, which means that they are harder to build mental models of and to understand. In addition, thinking itself is a complex system, if not *the most* complex system known to mankind. What we know about complex systems is that underlying their complexity are simple rules. Let's review the basic logic of the situation:

Table 2.4: Why Systems Thinking Must Be a CAS

If:	And:	Then:
Complex systems (CAS) are based on simple rules	Thinking systematically about complex systems is itself a complex system	What are the simple rules for systems thinking?

[8]2640 kcal/person/day of per capita food availability. 2015 World Hunger and Poverty Facts and Statistics. (2015, March 1). Retrieved from http://www.worldhunger.org/articles/Learn/world%20hunger%20facts%202002.htm

Complexity theory tells us to look for the simple rules that underlie the complex and adaptive human behavior that we know as "thinking." So what are the simple rules that underlie both the complex systems that make up the world and the way we think about those systems?

THE BIGGEST BARRIER TO SYSTEMS THINKING v2.0

As we have given talks and trainings throughout the world to diverse audiences, we have discovered one big barrier that impedes understanding systems thinking v2.0. Surprisingly, it isn't the simple rules themselves, which are actually remarkably easy to grasp. Instead, it is that many people hold the false assumption that *underneath complex things are complicated explanations.* The mental model they need is that *underneath complex things are simple rules.*

We've discovered that once people understand that simplicity underlies complexity, deep understanding and success in systems thinking v2.0 is a breeze. This is because understanding the simple rules is not difficult, but understanding how the simple rules work is not possible without understanding complexity.

Many people hold the false assumption that *underneath complex things are complicated explanations.* The mental model they need is that *underneath complex things are simple rules.*

COMPLICATED IS NOT THE SAME AS COMPLEX

For the last 2,500 years, our mindset has been that underneath complex systems there are probably layers and layers of complicated subsystems. Let me give you an example of what I mean by complex versus complicated. Kick a rock over and over again and it doesn't change its behavior. Each time, it follows the same laws of physics. The behavior of the rock is complicated but not complex; it doesn't adapt its behavior. It doesn't change. Now try the same experiment on a dog. Of course I don't mean literally go kick a dog; it's a thought experiment. Kick a dog and it recoils. Kick it again and it circles away. On the third kick the dog bites. The dog is a complex adaptive systems (or CAS), in that it adapts its behavior to better navigate its environment. In sum, a complicated system like the rock doesn't adapt, whereas a complex system like a dog adapts to survive in its environment.

COMPLEX ADAPTIVE SYSTEMS (CAS)

We now know that underlying complex adaptive systems are simple rules. Let's take a look at a video online to illustrate this as there's simply no way to experience it in writing. Go to any browser and enter the following URL:

Watch CAS Birds: crlab.us/stms

Take a look at the flocking behavior of what amounts to millions of starlings. Look at how quickly millions of birds pivot from all moving left to all moving right. It's called a *superorganism*, a bunch of individual organisms that act like a single organism. When scientists first began studying such systems we thought they must have exceptionally good leaders! These types of systems–seen across the physical, natural and social sciences in flocks, schools of fish, traffic patterns, ant colonies and across the spectrum of nature and human society–baffled scientists because it was unclear how the group behavior occurred in the absence of a leader. But there's simply not enough time for communication to occur between the leader and the follower, nor enough time for the signal to spread. What then causes this behavior?

It turns out there were no leaders, only followers. What were they following? They followed simple rules that brought about this remarkable, adaptive, and complex behavior. These types of systems are based on simple, local rules. Iain Couzin,[9] who studies collective animal behavior at Princeton University, did a simulation to show exactly what rules these flocks were following and found just three:

- Rule 1: maintain distance x (locally to nearest neighbors);
- Rule 2: adjust direction (locally to nearest neighbors); and
- Rule 3: avoid predators.

In the video you can actually see the simple rules perturb through the system as predatory hawks attempt to catch the birds. The birds at the bottom of the column are following rules one and two and have no idea that rule three (avoid predators) was followed above.

Figure 2.10: Example of CAS

9Couzin, I. D., Krause, J., James, R., Ruxton, G. D., & Franks, N. R. (2002). Collective Memory and Spatial Sorting in Animal Groups. *Journal of Theoretical Biology*, 218(1), 1-11. doi:10.1006/yjtbi.3065.

Humans do this, too. The largest human wave consists of 80,000 people acting as a single superorganism with no leadership, all following one simple rule: do what the person to your left does: when they stand, you stand. When they sit, you sit. Remarkable.

Watch the largest human wave on record: crlab.us/stms

There's a relatively simple formula for these complex adaptive systems: autonomous agents follow simple rules based on what's happening locally around them, the collective dynamics of which lead to the emergence of the complex behavior we see.

When we think in systems with our old mindset, we think like a field commander perched on a hill trying to design and control the complex behavior we want to occur. When we take a systems thinking v2.0 perspective, we think like individual soldiers and rely on the collective dynamics of the system to emerge. These collective dynamics produce emergent complexity (things like adaptivity, robustness, etc.). Simple rules underlie complex systems. Figure 2.11 depicts the basic features of complex adaptive systems.

Simplicity exists in many places in our everyday lives. The problem is, we distrust it. We don't think of simplicity as a good thing. When we think of someone intelligent, we think of someone who speaks in complicated ways, not someone who keeps things simple. When we face wicked problems, we don't think to ourselves the answer is simple. But the simplicity that underlies complexity is real. Here are a few examples of complex things that have relatively simple rules underneath.

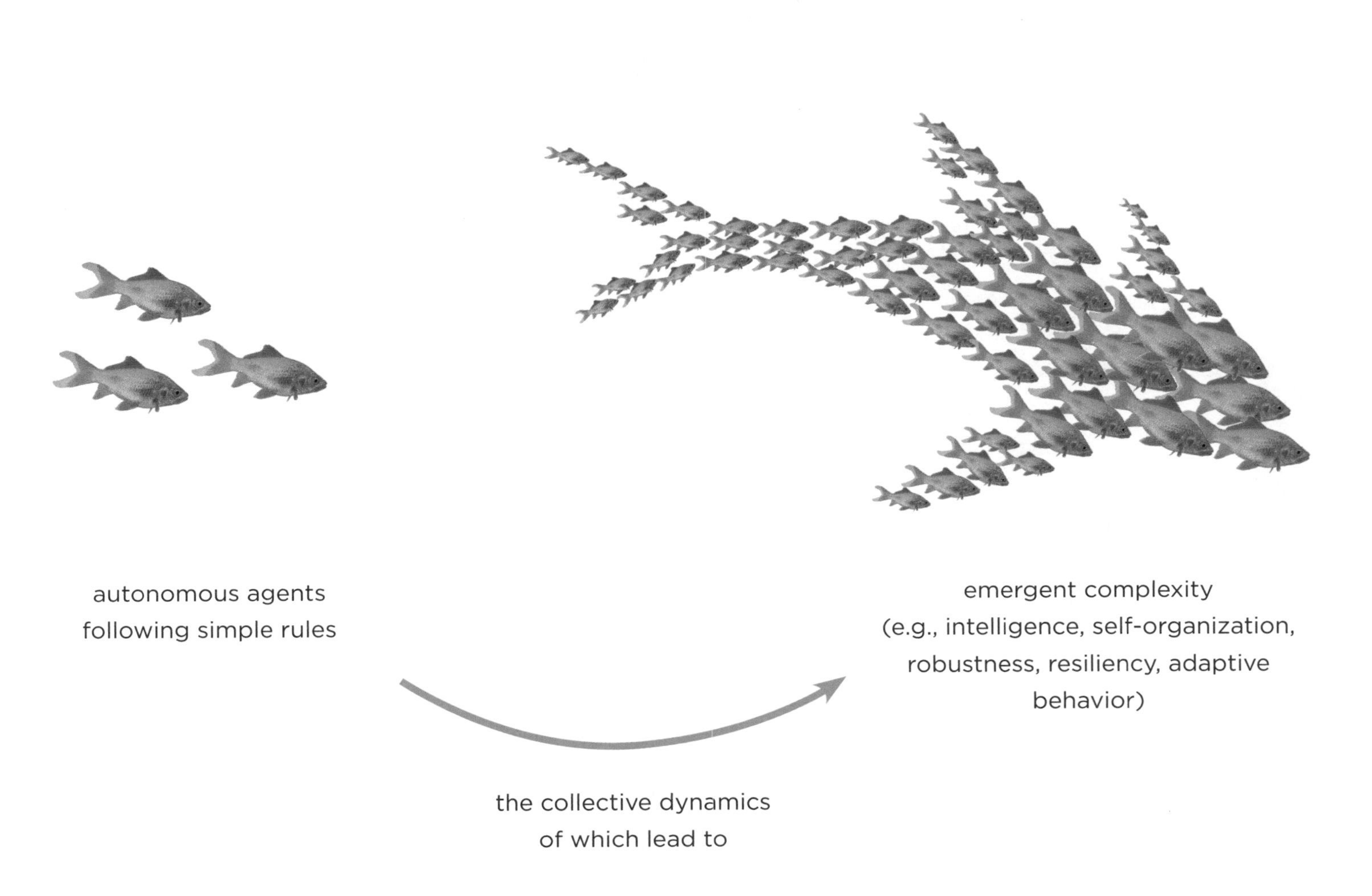

Figure 2.11: The Basic Features of a Complex Adaptive System (CAS)

RAINBOW OF COLORS: CMYK

Most of us have never seen the Mona Lisa, a Picasso, a Van Gogh, or a Michelangelo. A few have been lucky enough to see them in person at the Louvre or other famous places. But most of us have never seen the real thing. Yet we are all familiar with (and could describe) the Mona Lisa. How? Because of four simple colors—Cyan, Magenta, Yellow, and Black, or CMYK. What makes these four simple colors amazing isn't that they are the most beautiful; a world painted in only these four colors alone would be monotonous. What makes them special is that they interact together in a complex way that makes an infinite array of colors possible. This means that any of the great artists and paintings can be reproduced, or alternatively, that Home Depot can match your bathroom's Soulful Blue 6542 when your son rams the doorknob through the wall and you need to patch and repaint it.

Figure 2.12: Simplicity Underlies Complexity: Cyan, Magenta, Yellow and Black

BIODIVERSITY: ATCG

Likewise, few of us are awe-inspired by the mention of Adenine, Guanine, Cytosine, or Thymine, or ATCG, the organic molecules of DNA. But we sure do go bonkers when we see the beautiful biodiversity these four nucleotides can produce when mixed and matched by evolutionary processes. How do you look at a star-nosed mole rat or a seahorse or a giraffe or a platypus and not think: nature is really creative and has an amazing sense of humor.

Charles Darwin in *The Origin of Species* wrote, "from so simple a beginning endless forms most beautiful and most wonderful have been, and are being, evolved."[10] He only had an inkling of what we would later discover to be the hand of DNA, but his mindset was one of simplicity and complexity. Underlying all of the complex and adaptive biodiversity of life itself is simplicity.

Figure 2.13: Simplicity Underlies Complexity: DNA

[10] Darwin, C., & Mayr, E. (1964). *On the origin of species* (p. 490). Cambridge, Mass.: Harvard University Press.

LEARNING KARATE: WAX-ON/WAX-OFF

In one of our favorite scenes from *The Karate Kid*, Daniel-san has asked Mr. Miagi to teach him karate and all he seems to be doing is hard labor and household chores around Mr. Miagi's property. Daniel-san is pissed off and he's had enough and is going to quit, but he's about to have his mind blown instead. During the course of this conflict with Mr. Miagi, he realizes he has been taught four simple movements that underlie all of the strikes, kicks, and blocks that make up karate. Mr. Miagi is not only a Karate-master, he is also a master teacher. He knows that to achieve the outcome Daniel-san desires, his student will need to be able to adapt to the unpredictable future of bullying, fights, and competition. To prepare him for such an unpredictable future, rather than teaching him a laundry list of moves, Mr. Miagi focuses on the simple underlying fundamentals that can be combined and recombined: wax on-wax off, paint the fence, side to side, and sand the floor. These four simple rules combine in an infinite number of ways to develop Daniel-san as an adaptive, robust, and resilient fighter.

Watch Karate Kid: crlab.us/stms

Figure 2.14: Simplicity Underlies Complexity: Wax on, Wax Off

INFINITE BUILDS: MODULAR LEGO BRICKS

Ole Kirk Christiansen, a carpenter, founded Lego in 1932. At the time, he was out of work because of the Depression and decided to build wooden toys in Denmark. In 1947, Ole got samples of a plastic brick invented and patented ("self locking building bricks") by Mr. Hilary "Harry" Fisher Page in Britain, and began creating the automatic binding bricks which we know today as Lego Bricks, a name that originated in 1953. Ole's 1958 Lego patent (#3005282) states, "the principle object of the invention is to provide for a *vast variety of combinations* of the bricks for making toy structures of many different kinds and shapes." And that was the magic of Lego—vast variety from simplicity. Anything imaginable could be built. All kids could unleash their creativity on the world with simple, modular, relational blocks. Today Lego, with headquarters in Billund, Denmark, is the sixth largest toy company in the world, with over 5,000 employees and revenue of $7.8 billion Danish Kroner. Simplicity can be lucrative.

Figure 2.15: Simplicity Underlies Complexity: Simple Building Blocks

CHESS: PIECES AND MOVES

With just 6 unique pieces (sixteen total per side) and a simple set of local rules, perhaps nothing captures the complexity-simplicity paradigm better than chess. It is revered as a complex game, causing us to bestow grandmasters with the title of genius, yet it is a game a child can learn to play at an early age. Built on autonomous agents following simple local rules, chess's potential for complexity is immense.

The seemingly simple moves of chess combine into staggering mathematical possibilities. For example, there are 318 billion possible ways to play the first four moves. There are 1.7 x 10^{29} possibilities for the first 10 moves. The longest game of chess that is theoretically possible involves 5,949 moves. The longest actual chess game ended in a draw after 269 moves and 20 hours and 15 minutes of play.[11]

A chess grandmaster understands the simple rules that combine into numerous possibilities in the game of chess. The karate sensei understands the fundamental movements that combine to produce all karate moves. The systems thinker is not so different. The qualities, disposition, and skills that make up a systems thinker rely on a deep understanding of a simple set of rules. In the next chapter, you will see that—like bird flocks, biodiversity, karate sensei, or grandmasters—all systems thinking is predicated on simple rules.

simplicity underlies complexity

chess

six unique pieces

Figure 2.16: Simplicity Underlies Complexity: Chess

[11] Jindal, S. (2012, May 25). Interesting Chess Facts [web log post]. Retrieved from http://www.chess.com/blog/keshushivang/interesting-chess-facts2

SYSTEMS THINKING IS A COMPLEX EMERGENT PROPERTY OF FOUR SIMPLE RULES

With all these systems as examples, it shouldn't be hard to imagine that something as adaptive and complex as systems thinking could also be predicated upon simple rules. Systems thinking is not a process but an outcome—it is an ends, not a means. In other words, when we ask ourselves how we can become better systems thinkers it is important to realize that systems thinking is an emergent property. Systems thinking is a complex adaptive system. If we focus on what systems thinking is we will have little hope of actually achieving it. Instead we must focus on the simple rules and agents that bring about systems thinking. Figure 2.17 illustrates the basic idea behind all complex systems—that simple rules and agents lead to collective behavior and emergence. If systems thinking is an emergent property, then those aspiring to be better systems thinkers must focus their efforts where they have influence: executing the simple rules.

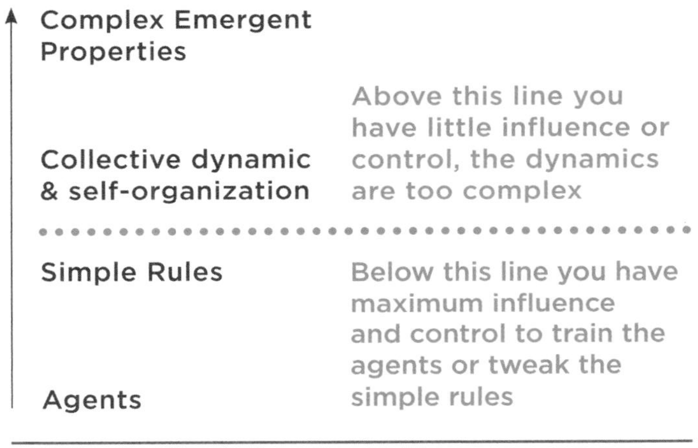

Figure 2.17: Where Systems Thinkers Should Focus Their Efforts

More than two decades of research went into discovering that there are four simple rules that underlie systems thinking which go by the acronym "DSRP:"

- Distinctions Rule: Any idea or thing can be distinguished from the other ideas or things it is with;
- Systems Rule: Any idea or thing can be split into parts or lumped into a whole;
- Relationships Rule: Any idea or thing can relate to other things or ideas; and
- Perspectives Rule: Any thing or idea can be the point or the view of a perspective.

One of the most important insights about DSRP is that it represents four cognitive functions that you must have just to form ideas. If your brain was rendered incapable of any one of them, it would have difficulty thinking about even the most basic things.

These four rules do not operate in isolation, but in parallel. A thing or idea can simultaneously be a distinct thing, a perspective, a part of a larger whole and a relationship. The next four pages provide a quick description for each of these four concepts as well as synonyms you will be familiar with. We'll further explain how they function dynamically as rules in the next chapter.

Figure 2.18: Simplicity Underlies Complexity: DSRP

thing

distinctions

other

Making distinctions between and among ideas. How we draw or define the boundaries of an idea or a system of ideas is an essential aspect of understanding them. Whenever we draw a boundary to define a thing, that same boundary defines what is not the thing (the "other"). Any boundary we make is a distinction between two fundamentally important elements: the *thing* (what is inside), and the *other* (what is outside). When we understand that all thoughts are bounded (comprised of distinct boundaries) we become aware that we focus on one thing at the expense of other things. Distinction-making simplifies our thinking, yet it also introduces biases that may go unchecked when the thinker is unaware. It is distinction-making that allows us to retrieve a coffee mug when asked, but it is also distinction-making that creates "us/them" concepts that lead to closed-mindedness, alienation, and even violence. Distinctions are a part of every thought-act or speech-act, as we do not form words without having formed distinctions first. Distinctions are at the root of the following words: compare, contrast, define, differentiate, name, label, is, is not, identity, recognize, identify, exist, existential, other, boundary, select, equals, does not equal, similar, different, same, opposite, us/them, thing, unit, not-thing, something, nothing, element, and the prefix a- (as in amoral).

Watch Systems Thinking Short: crlab.us/stms

systems whole part

Organizing ideas into systems of parts and wholes. Every thing or idea is a system because it contains parts. Every book contains paragraphs that contain words with letters, and letters are made up of ink strokes which are comprised of pixels made up of atoms. To construct or deconstruct meaning is to organize different ideas into part-whole configurations. A change in the way the ideas are organized leads to a change in meaning itself. Every system can become a part of some larger system. The process of thinking means that we must draw a distinction where we stop zooming in or zooming out. The act of thinking is defined by splitting things up or lumping them together. Nothing exists in isolation, but in systems of context. We can study the parts separated from the whole or the whole generalized from the parts, but in order to gain understanding of any system, we must do both in the end. Part-whole systems lie at the root of a number of terms that you will be familiar with: chunking, grouping, sorting, organizing, part-whole, categorizing, hierarchies, tree mapping, sets, clusters, together, apart, piece, combine, amalgamate, codify, systematize, taxonomy, classify, total sum, entirety, break down, take apart, deconstruct, collection, collective, assemble. Also included are most words starting with the prefix org- such as organization, organ, or organism.

relationships

action——reaction

Identifying relationships between and among ideas. We cannot understand much about any thing or idea, or system of things or ideas, without understanding the relationships between or among the ideas or systems. There are many important types of relationships: causal, correlation, feedback, inputs/outputs, influence, direct/indirect, etc. At the most fundamental level though, all types of relationships require that we consider two underlying elements: action and reaction, or the mutual effects of two or more things. Gaining an awareness of the numerous interrelationships around us forms an ecological ethos that connects us in an infinite network of interactions. Action-reaction relationships are not merely important to understanding physical systems, but are an essential metacognitive trait for understanding human social dynamics and the essential interplay between our thoughts (cognition), feelings (emotion), and motivations (conation). Relationships are at the root of many concepts: connect, interconnection, interaction, link, cause, effect, affect, rank, between, among, feedback, couple, associate, and join; most words with the prefixes inter-, intra-, extra-, such as interdisciplinary, intramural; most words with the the prefix co- as in correlate or cooperate or communicate; and types of relationships such as linear, nonlinear, causal, feedback, and mathematical operators such as +, -, /, x.

perspectives

point

view

Looking at ideas from different perspectives. When we draw the boundaries of a system, or distinguish one relationship from another, we are always doing so from a particular perspective. Sometimes these perspectives are so basic and so unconscious we are unaware of them, but they are always there. If we think about perspectives in a fundamental way, we can see that they are made up of two related elements: a point from which we are viewing and the thing or things that are in view. That's why perspectives are synonymous with a "point-of-view." Being aware of the perspectives we take (and equally important, do not take) is paramount to deeply understanding ourselves and the world around us. There is a saying that, "If you change the way you look at things, the things you look at change." Shift perspective and we transform the distinctions, relationships, and systems that we do and don't see. Perspectives lie at the root of: viewpoint, see, look, standpoint, framework, angle, interpretation, frame of reference, outlook, aspect, approach, frame of mind, empathy, compassion, negotiation, scale, mindset, stance, paradigm, worldview, bias, dispute, context, stereotypes, pro-social and emotional intelligence, compassion, negotiation, dispute resolution; and all pronouns such as he, she, it, I, me, my, her, him, us, and them.

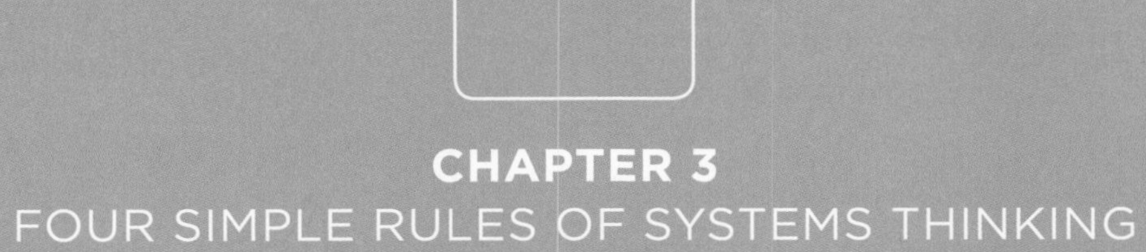

CHAPTER 3
FOUR SIMPLE RULES OF SYSTEMS THINKING

THE FOUR SIMPLE RULES OF SYSTEMS THINKING

A deeper understanding and greater application of systems thinking requires that we identify the patterns that connect all of the varied systems thinking methods. If all of these "big tent" methods and approaches (i.e., the MFS universe) are types of systems thinking, what cognitive patterns are universal to them all?

- Distinctions can be made between and among things or ideas;
- Things or ideas can be organized into part-whole systems;
- Relationships can be made between and among things or ideas; and
- Things or ideas can be looked at from the perspectives of other things or ideas.

These cognitive patterns are not merely applicable to systems thinking, they are universal to all thought. We will alternately refer to them as patterns, simple rules, and cognitive structures. Stated as simple rules, the patterns are:

- Distinctions Rule: Any idea or thing can be distinguished from the other ideas or things it is with;
- Systems Rule: Any idea or thing can be split into parts or lumped into a whole;
- Relationships Rule: Any idea or thing can relate to other things or ideas; and
- Perspectives Rule: Any thing or idea can be the point or the view of a perspective.

An important feature of each rule (DSRP) is that it involves co-implication. For example, the existence of a thing automatically implies the existence of an other, and vice versa. The same is true for the Systems, Relationships, and Perspectives rules. Therefore a part implies the existence of a whole, an action a reaction, and a point (looker) a view (looked at). The four DSRP Rules (or patterns) and their co-implying elements are systems thinking. Successful systems thinkers commit them (Table 3.1) to memory.

Table 3.1: 4 Patterns and 8 Elements

Simple Rule or Pattern	Element 1	Element 2
Distinction (D)	thing (t)	other (o)
System (S)	part (p)	whole (w)
Relationship (R)	action (a)	reaction (r)
Perspective (P)	point (ρ)	view (v)

While the simple rules of systems thinking are DSRP, the agents are little bits of information. *Systemic thought* emerges from bits of *information* following *simple rules* (DSRP). At first, it might seem strange to impute agency to information. Yet we all know that information seeps into our heads and often ideas pop out without our conscious choice. We also know from scientific disciplines such as neuromarketing (the combination of neuroscience and marketing) that it is not only possible, but advantageous for companies to use their understanding of the human mind to subconsciously manipulate our thoughts and cause us to make buying decisions. Likewise, neuromarketers are used in high-stakes political elections and debates.

Also remember that we interact with the world indirectly through our mental models, not directly. When one considers this fact, the thing (real world) and the idea (mental model) that represent it are effectively the same. So in our descriptions we will use the term idea and thing interchangeably. The mental model of a thing is determined by both its information and its structure—the simple rules it follows (DSRP).

The complexity and adaptivity that makes a CAS unique and robust emerges out of many autonomous agents (e.g., birds, fish, people) following simple rules. Now imagine that instead of birds, fish, or people, the agents in a system are *bits of information*.

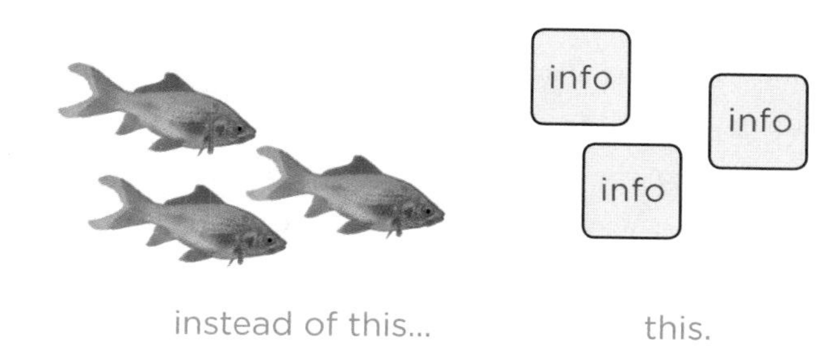

Figure 3.1: Information as Agents

When we take a CAS perspective on systems thinking, we ask ourselves: what are its simple underlying rules? The simple rules are based on distinctions (D), systems (S), relationships (R), and perspectives (P). That is, each bit of information can distinguish itself from other bits, each bit can contain other bits or be part of a larger bit, each bit can relate to other bits of information, and each bit of information can be looked at from the perspective of another bit of information and can also be a perspective on any other bit.

DSRP RULES OCCUR SIMULTANEOUSLY

It is important to realize that while a bit of information might be a distinction, it can simultaneously be a part-whole system, a relationship, and a perspective. Neither D, S, R, nor P exist in isolation; they co-occur. Thus, if we think of any bit of information as being represented by a simple square, then we could use the four corners of the square to represent the four rules that the square (a bit of information) could follow. We can use the four corners and four colors (Figure 3.2) to represent DSRP rules. Note that the bottom right corner of the square reads "action-reaction relationships." All types of relationships—correlational, causal, feedback, etc.—involve an action and a reaction.

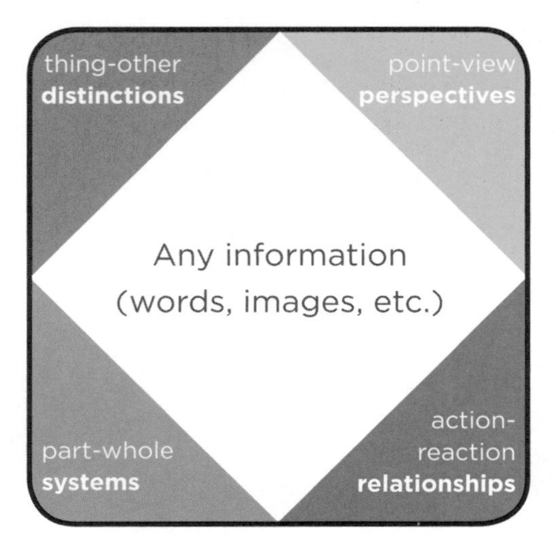

Figure 3.2: Information Follows Four Simple Rules

The images here are provided by a software program our research lab developed. The software, called *MetaMap*, treats every bit of information (text or image) as a rule-following square. Each square follows DSRP rules, which can be executed one at a time or simultaneously. Each square has four corners (see Figure 3.2) that light up when you mouse over them. The user decides which rule(s) to execute for that square. These little rule-following squares allow us to *visualize systems thinking*. So instead of this:

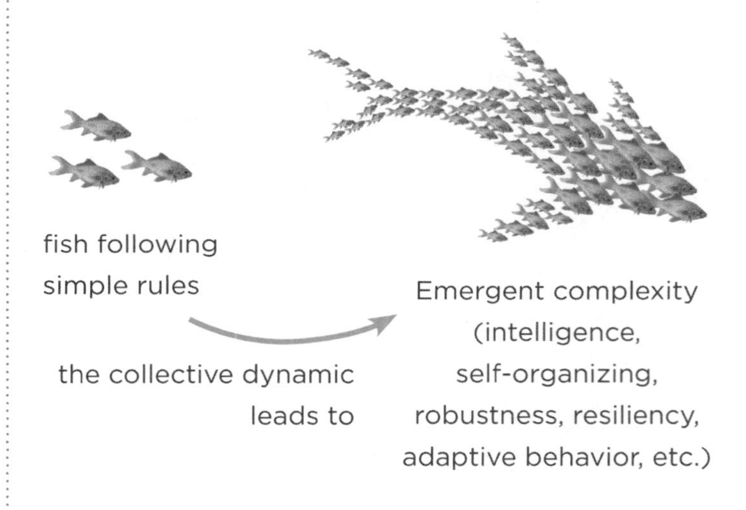

Figure 3.3: Fish and Simple Rules Lead to Superorganism

We are talking about this:

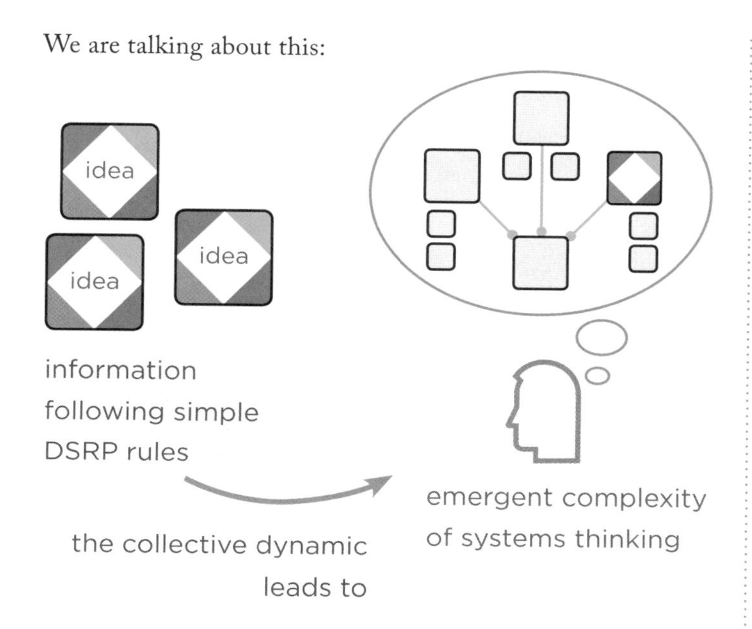

Figure 3.4: Ideas and Simple Rules Lead to Systems Thinking

Four simple rules produce collective dynamics that in turn emerge as systemic thought. This is quite simple, but it is also complex. Although DSRP rules are simple, the result of them applied over and over again is massively complex. Let's take a quick look at how each of these simple rules works and what it looks like using the abstract notation of squares to represent ideas or things.

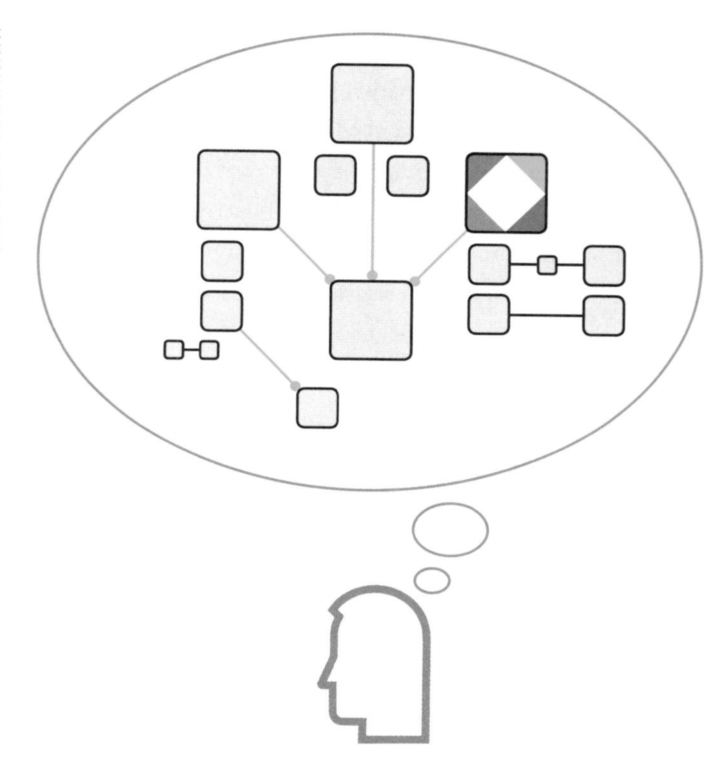

Figure 3.5: Every Object in a MetaMap has the Potential to Follow DSRP Rules

Four simple rules produce collective dynamics that in turn emerge as systemic thought.

To review, DSRP rules operate on information simultaneously and a single bit of information can be a distinction, system, relationship, and/or perspective. Imagine, for example, a systemic diagram representing some set of ideas or a network as in the thought bubble in Figure 3.5. First, notice that each bit of information in the network is distinct from other bits (squares). Note that when a relationship is distinguished, a square exists in the center of the line indicating not only that there is a relationship, but also explicating what it is. Some of the relationships (lines) have not yet become distinctions (i.e., they are currently undefined). Some of the squares are also whole systems because they contain parts, whereas other squares are not yet whole systems (perhaps because we haven't explored them yet). And some but not all of the squares are acting as perspectives, viewing or experiencing the system in different ways from each other.

DISTINCTIONS RULE:
Any Idea or Thing Can Be Distinguished from the Other Ideas or Things It Is with

What makes something a distinction is not so simple a question. Every thing or idea has an other. But in many cases the other is either implicit or absent in your thinking. If a thing or idea exists, then an other exists, even if it's not clear what the other is.[1] Let's say that we wanted to distinguish what a teacup is. We can define a teacup in two ways. We can describe to you every possible detail of the cup, its structure and function, patterns and meaning, *or* we can describe everything in the universe *except* the tea cup.

Figure 3.6: There Are Two Ways to Define a Tea Cup

Often we understand what ideas or things are by what they are not. This is true for all ideas, things, and systems. They all have boundaries that distinguish what's in and what's out and those boundaries are based on others, because a boundary is shared with another. The things we see and think about derive meaning from other proximate things and ideas. You could think of this as context and we often say that context changes everything. What's important to realize is that context for a thing or idea isn't a mysterious cloud that surrounds it, it's other things and ideas.

[1] In earlier work we referred to the thing or idea in the thing-other distinction as the "identity."

The things we see and think about derive meaning from other proximate things or ideas. You could think of this as context and we often say that context changes everything. What's important to realize is that context for a thing or idea isn't a mysterious cloud that surrounds it, it's other things or ideas.

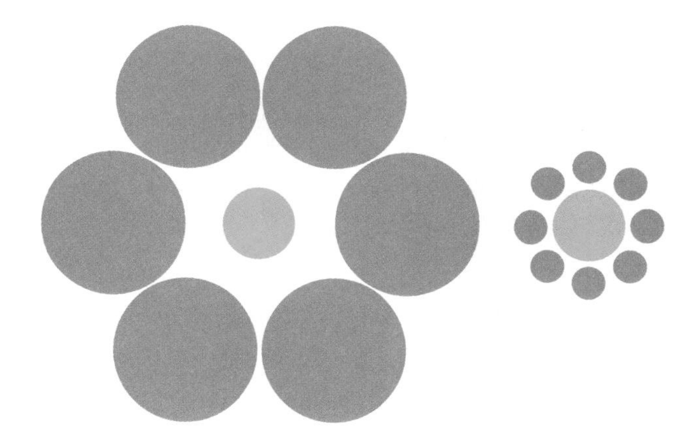

Figure 3.7: Things Are Defined by the Things That Surround them

That's precisely why, despite the fact that the two orange circles in Figure 3.7 are exactly the same size, they appear to be different sizes. Your mind is distinguishing them in relation to what is near them. You can see that the *other* is often a whole system or set of things (e.g., all of the gray circles surrounding the orange circle). Also important to note is that while each circle is distinct from the others, there is also a distinction between the *group* of circles in A and those in B. At every level of scale we are making distinctions, boundaries between what something is and what it is not.

But let's try to represent this rule in an abstract way so that we can use it over and over again for any idea or thing. In Figure 3.8, each idea is represented by a square, such that it is clear that A, B, and C are distinguished from one another because they are three different squares. If we say A is the thing we are distinguishing (like a teacup), then B and C are not A, or everything but A (like the not-teacup). A is the *thing*, and B and C are the *other* things. Note also (Figure 3.9) that we can define A in terms of how it relates to B and C. So it becomes "not-B or -C." Alternatively, B and C are "not -A."

Every idea starts with a distinction. Even the simplest thought involves drawing a boundary that distinguishes something from nothing or a thing from other things. Most of the time we communicate these ideas with words, yet words fail to communicate the hidden elements of our thoughts. This is true of all the terminology we use, but especially terms that are political or divisive, such as torture, terrorist, Muslim, conservative, liberal, us, or them. These words mean what we make them mean and include what we choose to include in their meaning. What these words mean is determined by what we load into them and what we don't. Another example is that for some people the concept of Sponge Bob may contain within it the degradation of the intellect and the decay of the fabric of society, whereas for others, it's just a funny character who is part of a kid's show. Any idea or thing that we might represent with words—dog, socialism, run, it, Sponge Bob, or any other of the over one million words in the English language—defines not only what something is, but what it is not.

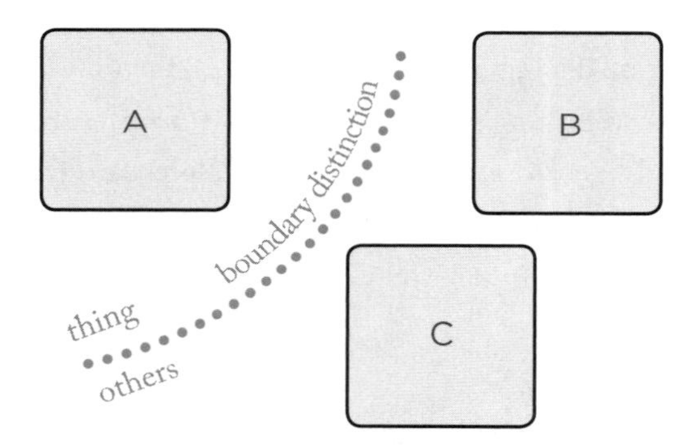

Figure 3.8: Distinctions Are Boundaries, Not Things

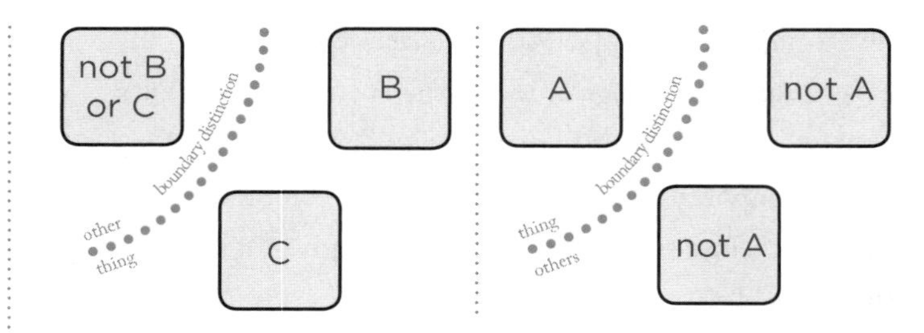

Figure 3.9: Distinctions Operate in Both Directions

The combination of dictionary and thesaurus is really a complex (and adaptive!) system of distinctions that includes definitions that are interlinked with other definitions and supported by words that are both similar (synonyms) and different (antonyms). On top of all that, these definitions, antonyms, and synonyms are not static, they're dynamic, contextual, and personal. The meaning of a word varies by context, by the other words used to modify it, and by the intent or ignorance of the user. What is important is that our distinctions are actually distinct. This may seem obvious, but it is frequently not the case. One of the first places that distinction errors can occur is in language.

Specifically, there are two ways that distinctions end up not being very distinct:

1. We use the same words to describe things or ideas that are different, i.e., a semantic problem or language error; and
2. We use different words to describe things or ideas that are actually the same.

To ensure that our language is not impeding our distinction-making, we must examine our conceptualization of distinctions. An advanced tool we can use when considering systems of distinctions is the mnemonic MECE (or NONG).

- MECE: Mutually Exclusive and Collectively Exhaustive
- NONG: No Overlaps, No Gaps

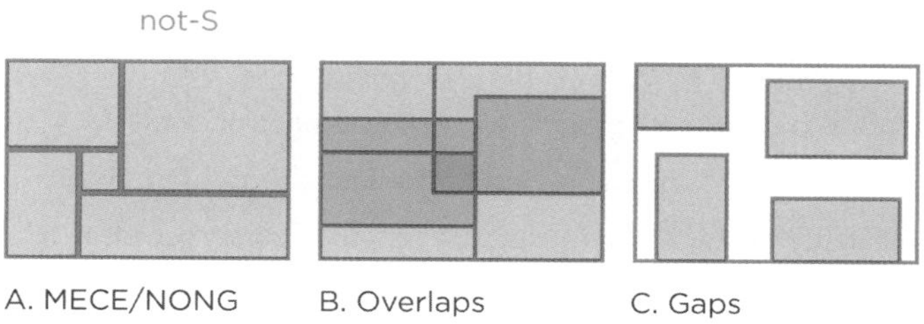

not-S

A. MECE/NONG B. Overlaps C. Gaps

Figure 3.10: MECE/NONG

Figure 3.10 visually depicts the MECE and NONG requirement filled (A) and not filled (B and C). MECE and NONG mean the same thing, so you can use whichever one makes more sense to you. No overlaps and mutually exclusive mean that the distinctions you make in a system are not overlapping. They are in fact distinct. No gaps and collectively exhaustive mean that the system of distinctions you have assembled to describe the problem is sufficient and complete and that everything that needs to be considered has been. MECE and NONG establish what is inside and outside the boundary of any system. What is important to understand is that while MECE and NONG help you to consider the system of distinctions that you are using, all of the items inside your system of distinctions is also a thing-other distinction. Distinctions are occurring at different scales with regard to the smallest ideas and things as well as the largest ideas and things. So you can see that in (A) there are distinctions occurring wherever there are red lines, which includes not only the distinctions between the parts inside the system, but also the distinction between what lies inside and outside of the system (not-S).

In solving wicked problems, one of the most important steps is identifying what the problem is and what it is not, what is inside the boundary of the problem and what lies outside. When our solutions fail us, upon tracing our steps backwards, we often find that the definition of the problem itself (i.e., the distinctions we made at the outset) were flawed.[2] Boundary critique (i.e., making one's distinctions distinct) is the process by which we test our distinctions. Foundational to this process is recognizing that every time we establish the identity of a thing or system of things, we also need to assess what that means for the other. For example, if we say that the thing is A,

---

[2] Rochefort, D. A. & Cobb, R. W. (1994). *The Politics of Problem Definition: Shaping the Policy Agenda.* Lawrence, Kansas: University Press of Kansas.

then we should ask what is the other, what is not -A? If we do not, we are at risk of bias that will come back to haunt us, or worse, marginalize the other, which can generate what we call violent distinctions.

Let us give you an example of a violent distinction. Let's say you go to your parent-teacher conference where the teachers express their concern that your child is disconnected and may need a professional learning assessment. Surprised, you quickly flash through your child's life looking for first-hand experience of a disconnected child, to no avail. The comment doesn't resonate because it doesn't match your experience. So you inquire more: what does disconnected look like? You learn that at seven your child is not engaged in the classroom work and distracted. In other words, he is connected to something else that is more interesting to him. This is a violent distinction. The child is not disconnected, he's differently connected. To say that he is disconnected is to imply that he is not connecting to anything, which would be a problem. But to define the problem as disconnection, when the child is highly engaged in the things that interest him, and then to recommend professional (and often stigmatizing) assessment can be seen as a violent act, as it has the potential to pathologize childhood. We label far too many normal children with stigmatizing labels like attention deficit disorder, yet in many cases it could just as easily be called boredom intolerance. If the thing is disconnection, then the other is connection. But he is capable of connection, therefore the root problem is not disconnection, but a mismatch between what the child is connected to and what the teacher wants the child to be connected to. Violent distinctions can be subtle like a simple word choice in a teacher conference or they can be egregious and explicit like any of the many us-them distinctions that are drawn in world politics, religion, and society.

SYSTEMS RULE:
Any Idea or Thing Can Be Split into Parts or Lumped into a Whole

There's an old saying that there are really only two types of scientists in this world, splitters and lumpers: those who tend to break things into parts (deconstruction) and those who tend to put parts together into a whole (construction). We need more "splumpers:" people who have the ability to both construct or synthesize ideas, and also to deconstruct ideas to further our understanding. This is a key idea in systems thinking and it is part and parcel of being a systems thinker.

The second rule means that all ideas and things can be either broken into smaller things and ideas or lumped into larger ones: part-whole. Remember, holism without partism, and vice versa, is meaningless. We need to consider both.

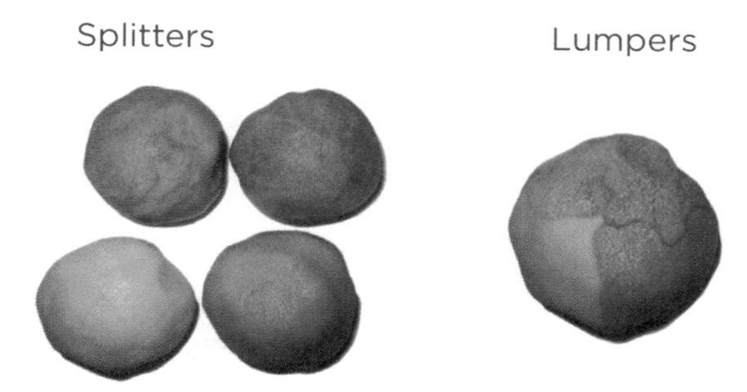

Figure 3.11: Splumpers Needed

How might this be communicated visually in an abstract way so that it could apply to and be used for any idea, no matter what the topic or system being analyzed? Returning to our representational square (Figure 3.2) we see that a natural metaphor is that of containment. So we simply show that the parts are contained by the whole. But visually when we put the parts inside the whole it can get crowded and become messy, so in the same way that we do for a traditional outline, we can think of the parts as coming out of the whole and organize them below. Later you will see that we can choose ways to organize them other than an outline format. The whole (a square node) is always larger than the parts (a smaller square node).

In Figure 3.12 we see that everything in the diagram is already a distinct thing (A, B, C, B1, B2, C1, and C2) because they are all represented by a square. But some are wholes that contain parts (A, B, and C), some are wholes and parts at the same time (B and C), and some are just parts (B1, B2, C1, and C2). Of course, at times we want to see the complexity of all the parts and at times we don't, so a simple toggle (▾) can be used to show or hide the parts.

Remember, what makes something a part is that it belongs to a whole. And what makes something a whole? It has a part. Part implies whole and whole implies part; they co-imply each other. The systems rule guides us first to think about a thing as a whole and then to consider what the parts might be. It also asks us to consider what that whole might belong to (i.e., what is it a part of?). Every whole has the potential to be a part, and vice versa, but your mind needs to do the work to see this. In the real world, whatever you're looking at has parts. Figure 3.11 illustrates that each time we think that we have discovered the smallest thing (i.e., something that doesn't have parts or is irreducible) or the largest thing (i.e, something that is not part of a larger whole), we soon learn that we're wrong. The term *atom* means "not splittable" (a-tom). Figure 3.11 shows that the atom exists around 10^{-11},which means that since the discovery and naming of the atom, we have discovered many smaller-scale things. On the other end of the spectrum of scale, the *Universe* was likewise named for its singular existence ("uni-" means one). At 10^{-24} on Figure 3.11, it was the whole that was part of nothing else. Today, we know that there's quite a bit more out there (i.e., the Universe is part of a meta-verse made up of many universes).

A is just a whole

A

B and C act as parts of A but are also wholes that contain parts

B

B1

B2

B1 and B2 as well as C1 and C2 are just parts, but they have the potential to be wholes

C

C1

C2

Figure 3.12: Part-Whole Diagram

To reiterate, the systems rule tells us that any idea or thing can be a whole or a part. It also shows us how a bunch of things or ideas can be organized into part-whole groupings (Figure 3.13).

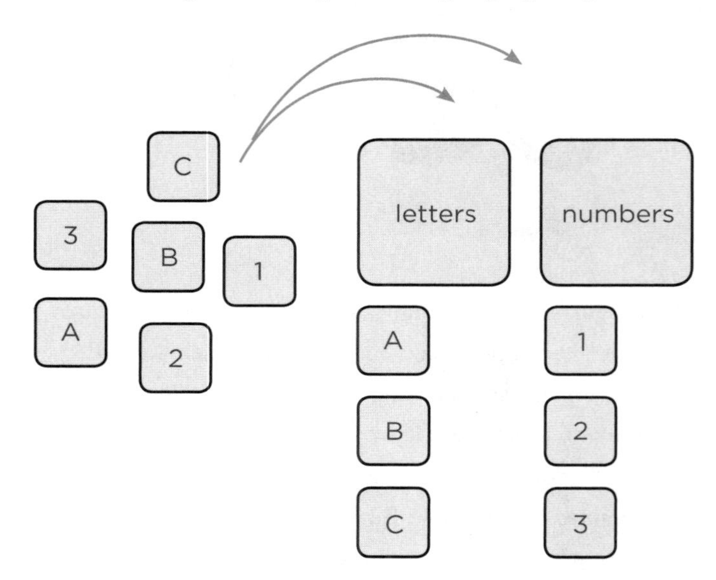

Figure 3.13: Organizing by Part-Whole

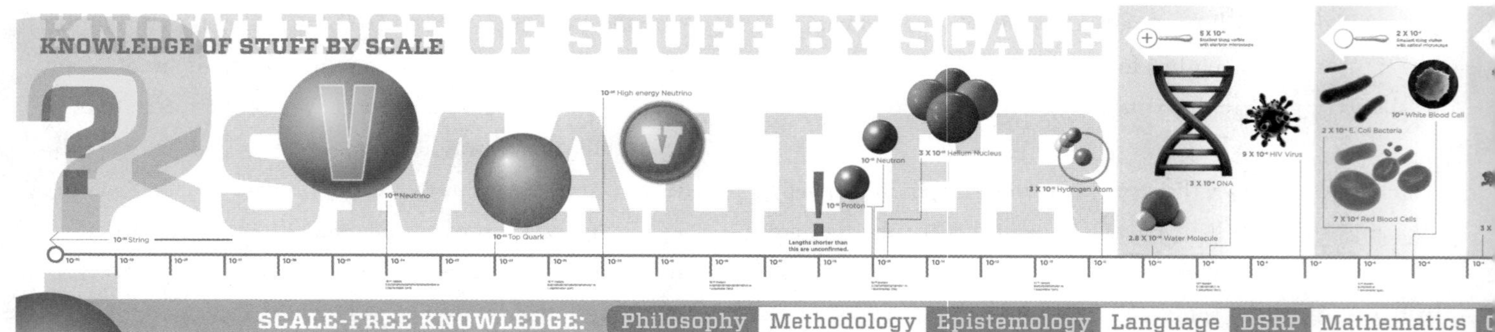

RELATIONSHIPS RULE:
Any Idea or Thing Can Relate to Other Things or Ideas

Relationships (e.g., connections, links, feedback, causality, likes, and friendships) are ubiquitous across all systems. Note that a relationship can be physical and tangible (such as a cord is the relationship between my laptop and electricity) or physical but somewhat invisible (the relationship between a magnet and iron particles) or conceptual (the relationship between war and peace).

We are quite accustomed to drawing relationships between and among ideas or things. Most of us simply draw a line, and so this is a simple but powerful way to visualize relationships. Although this may seem drop-dead simple (and it is), remember that the four simple rules can combine. This allows us to do things to this simple line (e.g., distinguish it as an idea or convert it into a whole with parts) that will change everything about how we think about systems. But for now, a simple line will do. We programmed the little idea nodes so that when you drag the relationship corner from one idea node to another, it forms a line. Of course, without the programmable blocks, you can do the same in your mind, or on a sheet of paper. We can also add arrows or multiple relationships to show directionality of relationships between and among things. Figure 3.14 illustrates the basic things we can do with relationships and directionality.

Watch Exploding House: Be sure to watch this amazing Lowe's commercial where a house "explodes" into parts and parts of parts... and then comes back together into a whole, and wholes of wholes. All systems have these properties. crlab.us/stms

Figure 3.14: Part-Whole Structure in Everything

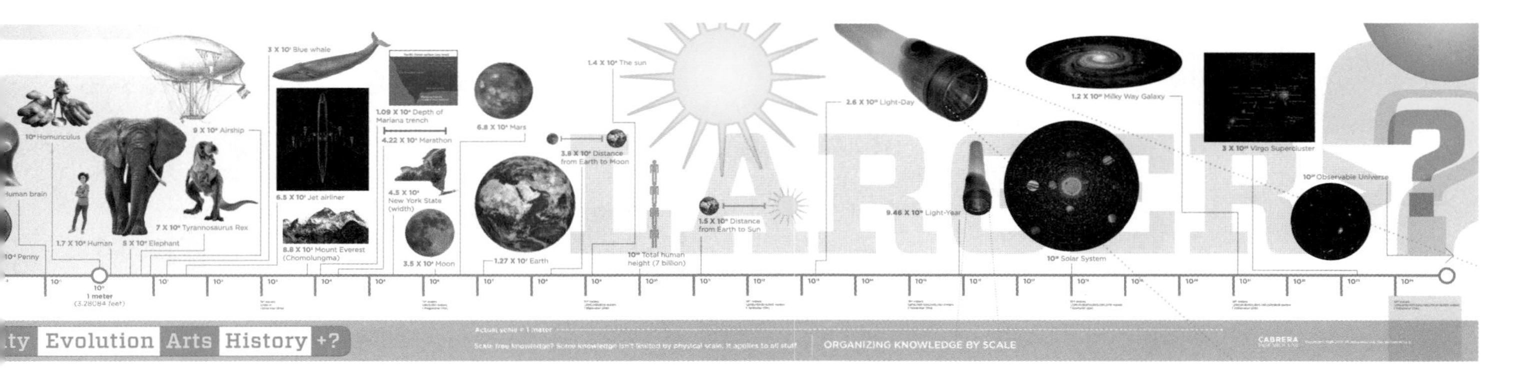

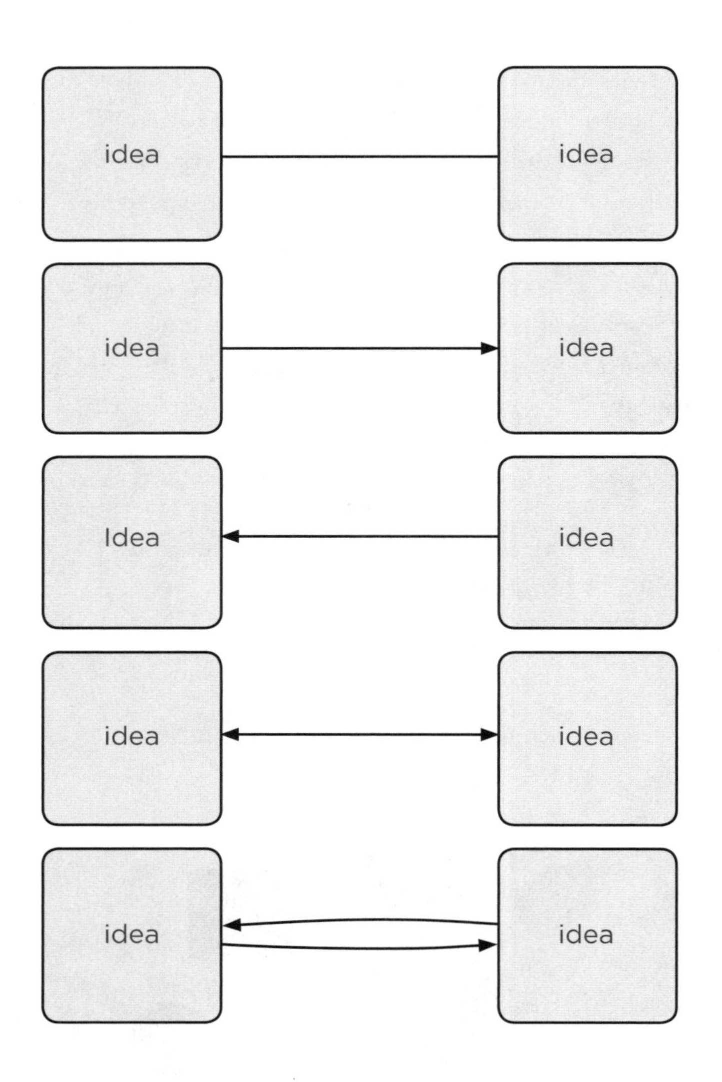

Figure 3.15: Lines as Relationships, Arrows as Directionality

PERSPECTIVES RULE:
Any Thing or Idea Can Be the Point or the View of a Perspective

In systems thinking, perspectives are as ubiquitous as relationships. In many cases, it is the various perspectives that factor into complex systems and their wicked problems that create the need for systems thinking. Most of the time, we think of a perspective as being something that a person has (i.e., "his or her perspective"), but a perspective can also be a lens through which we look at some other idea and that lens can be a thing, an event, a person, a place, and even an idea. Whether you call it a lens, a point of view, a frame, mindset, or worldview, all are synonymous with or forms of perspective-taking. Perspective-taking is essential to systems thinking because understanding systems is in large part understanding how those systems look from various perspectives.

In the same way that Figure 3.15 illustrates that two angles on the same object can yield very different perspectives, we intuitively understand that the same is true about all things and ideas. But perspectives are not irreducible, they are comprised of two lesser elements: a point and a view.

In order to capture this abstract underlying structure, we programmed these little idea nodes to be able to become a point or a view (or in some cases both) so that we can easily visualize perspectives. You can see in Figure 3.16 that there are two perspectives that are each made up of a point and a view. In this case, the view is the same for both perspectives (idea C). Any idea or thing or even a whole system (i.e., any square) can be a point, or a view, or both.

Figure 3.16: Two Points Yield Different Views of the Same Object

In many cases, it is the various perspectives that factor into complex systems and their wicked problems that create the need for systems thinking.

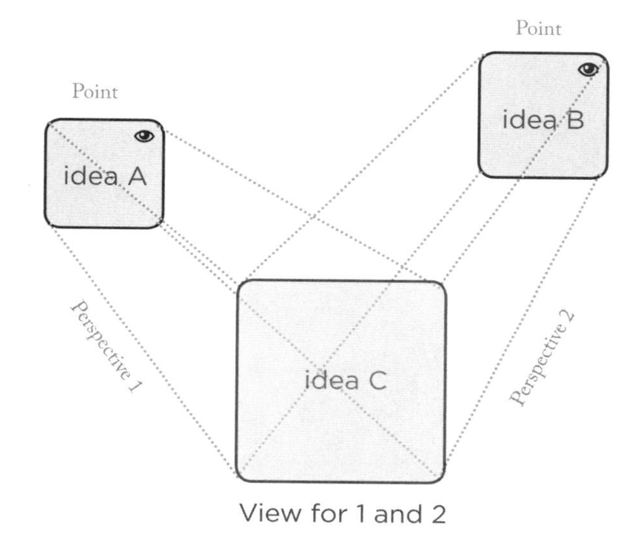

Figure 3.17: Different Perspectives Result from Changing the Point, the View, or Both

Creating all the lines for a perspective isn't necessary and adds visual clutter. Therefore, we reduce this basic abstract principle of perspective by showing a perspectival line (orange) and differentiating the point from the view with an eye icon (👁). In Figure 3.17, we see three different points looking at the same view (i.e., three perspectives).

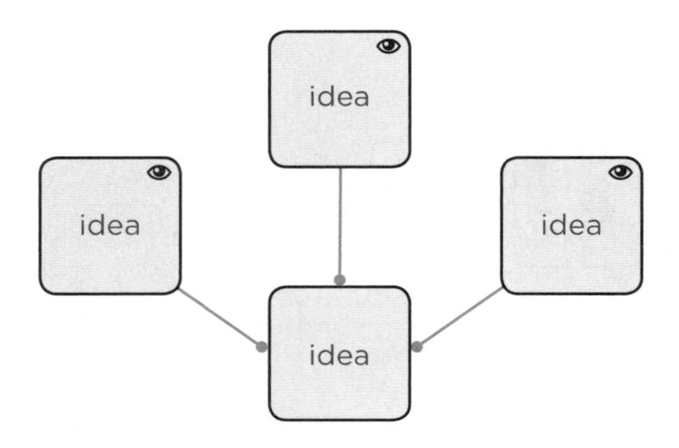

Figure 3.18: Three Perspectives on One Idea

Perspectives can be used to make us expand our thinking and include more options (i.e., divergent thinking). It can also be used to restrict our thinking and cause greater focus (i.e., convergent thinking). You could say, for example, "let's look at this topic from the perspective of X, and Y, and Z" and therefore cause thinking to expand and include more ideas. In contrast, you could also say, "let's only look at this topic from the perspective of X" and therefore occlude other perspectives on the topic. This allows us to focus on the salient issues at hand when necessary. This can be particularly useful in facilitating team meetings or classrooms where the facilitator needs to keep people focused and on task but also avoid giving the impression that divergent thinking isn't valued: "There are many other important perspectives on this topic, but today we want to focus on only this perspective."

SYSTEMS THINKING MADE SIMPLE

Now you can see that what we mean when we named this book "Systems Thinking Made Simple" is not to communicate that we are simplifying it per se. Instead we are showing that systems thinking is actually a complex, adaptive, and emergent phenomenon that is based on simple rules: DSRP. It just requires a little practice to get your brain in the habit. For many, we have found that the best way to do that is by visually mapping mental models, which we'll explore more thoroughly in the next section. It will also help to understand that while these rules are simple, their implications can be quite surprising. In Chapter 5, "Use And Reuse Cognitive Jigs," you'll see that common structures emerge out of systems thinking; knowing these structures will make you a faster and better systems thinker and help you solve problems in your topical areas of concern. In chapter 6, "Make Structural Predictions," you'll learn that DSRP can be used to discover new knowledge in any discipline, while chapter 7, "Embrace And/Both Logic," demonstrates how DSRP simple rules underlie an entirely new and powerful logic. In chapter 8, "Everyday and Advanced Applications of DSRP," you'll see that DSRP simple rules underlie all other systems modeling techniques. Finally, in section 3 of the book, "7 Billion Systems Thinkers," you'll learn both how DSRP can lead to a change in the way we think about personal development, education, and employee training, as well as how these rules can transform organizations through systems leadership.

SECTION 2

BECOMING A SYSTEMS THINKER

SECTION 2
BECOMING A SYSTEMS THINKER

Becoming a systems thinker takes practice. What we have found in our research on public understanding of the field is that those who develop a deep understanding institute a number of changes, both general and specific. On the whole, systems thinkers we have taught do the following: practice, do not fear making mistakes or doing it wrong, and learn to metamap.

Practice is important, because learning something new, especially when it is something paradigmatic like systems thinking, has as much to do with developing deep understanding as it does with burning neural pathways. Neural pathways are like trails in the woods, if we walk them often they become well trodden, what was once obscured by underbrush becomes more and more like a path, and the way becomes clearer and the walking faster. This is also true of our thought processes. Fear plays a key role just like it does in learning any new language. Fear of making mistakes can make a foreigner apprehensive to speak the language and, in turn, cause them to speak it less and therefore get less practice, less immediate feedback, and less progress in learning the language. Newcomers struggle with the fear that they are not doing it "right." Finally, although systems thinking occurs in the mind, as will increasingly be the case as expertise develops, visually modeling one's ideas using metamaps has proven invaluable not only in understanding and communicating one's systemic ideas, but also in learning systems thinking. Metamapping is not the same as applying DSRP rules, because there are obvious limitations to the flatland of visual maps. What the mind can do with DSRP will always be more complex and robust than what can be done on paper or in software. Yet, metamapping makes the essential processes of systems thinking tangible and therefore is a useful tool for learning.

We have also seen that budding systems thinkers do a few more specific things on their way to developing expertise. This section focuses on the general actions above as well as these more specific actions in the chapters that follow..

Chapter 4
See Information and Structure

They understand that mental models are made up of information content and structure (DSRP simple rules), and pay as much attention to the structural details as the informational ones. They utilize visual and sometimes tactile tools to see this structure and they understand how the cognitive style of some visual mapping techniques hinders systems thinking.

Chapter 5
Use and Reuse Cognitive Jigs

They use common patterns in the structure of systemic mental models, called jigs, to gain insight and increase the efficiency and speed of thought.

Chapter 6
Make Structural Predictions

They use the structural properties of mental models (DSRP simple rules) to make predictions that lead to new understanding and discoveries.

Chapter 7
Embrace And/Both Logic

They use the structure of DSRP simple rules to embrace multivalent (and/both) logic and avoid the pitfalls and biases of binary (either/or) logic.

Chapter 8
Everyday and Advanced Applications Of DSRP

They apply DSRP to everyday tools like tables and graphs, but also to complex methods such as network theory and system dynamics.

CHAPTER 4

SEE INFORMATION AND STRUCTURE

VISUALIZE INFORMATION AND STRUCTURE

Systems thinking diagrams called "metacognitive maps" or "metamaps,"[1] facilitate learning systems thinking by making its simple rules (DSRP) explicit. In fact, metamaps and thinkblocks are used with equal success to teach preschoolers, Ph.D.s, CEOs, and engineers how to increase their systems thinking abilities.

Using metamaps to visualize systems thinking[2] shows us that seeing is understanding. Whether your desire is to capture, clarify, or communicate your ideas, learning to visually map systems thinking in this way will allow you to:

1. Gain deeper understanding of systems thinking and the systems you're thinking about; and
2. Visually capture mental models and communicate to others.

It is easy to conclude that when we read or speak it is the words that carry meaning. But there is a hidden structure that underlies words, which linguists call grammar and syntax. To understand how this structure contributes to meaning, take a look at the following sentences:

Woman, without her man, is helpless.
Woman! Without her, man is helpless!

I like to eat Grandma.
I like to eat, Grandma.

Notice that while the information content (the text) is the same for both sentence pairs, the meaning is quite different. If the text is the same and the meaning is different, then there must be some other factor—the grammatical structure of the sentence—contributing to the meaning.

In systems thinking, the equivalent of text is *information* of any kind, including words, numbers, symbols, labels, and data. The equivalent of grammar and syntax is *structure.*

Based on this distinction between information and structure, let's develop a common set of terms to better understand that hidden structure—not just information—is critical to becoming a systems thinker. Table 4.1 illustrates the importance of information and structure to understanding (and systems thinking).

---

[1] The term metamap(s) is used to refer to any kind of metacognitive map (pencil and paper drawing, one produced in the software, etc.). When referring to the metamapping software we use the capitalized spelling MetaMap.

[2] The point of learning to metamap is not that you use these maps exclusively. For example, if you're dealing with population models or inventory models with lots of feedback, then there are software programs that help you do that and quantify the items in your map. Nonetheless, those maps and any other systems thinking methods of mapping are using the four simple rules of systems thinking. You'll see numerous examples of this in the entitled Everyday and Advanced Applications Of DSRP.

Table 4.1: Information and Structure = Mental Model

Information		Structure		Mental Model
Includes all material, information, or data of any kind, including written text or spoken words and images that contribute to meaning.	+	Includes the hidden contextual structure that contributes to meaning.	=	Is the essential concept or meaning. The purpose of any communication is to relay meaning of one's mental models.

The problem is that most of the communication techniques we rely upon—including not only the spoken and written word, but also the various mapping and modeling techniques we use—are focused heavily or exclusively on information. Because the structure of the mental model is overlooked, meaning is often the casualty of presentations and communications.[3] Becoming a systems thinker entails becoming someone who sees both the information and the structure.

Metacognitive maps highlight the underlying structure or grammar of thought; rather than merely the information—a core skill in the process of becoming a systems thinker. In contrast, most of the existing visualization techniques available to systems thinkers are "informational maps" that give greater weight to information content (e.g., words).

[3]Tufte, E. (2006). *The Cognitive Style of PowerPoint: Pitching Out Corrupts Within*. Cheshire, Connecticut: Graphics Press.

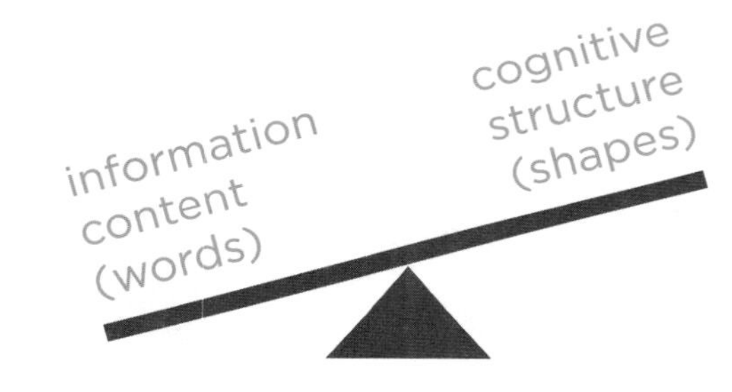

Figure 4.1: Informational Maps Are Out of Balance

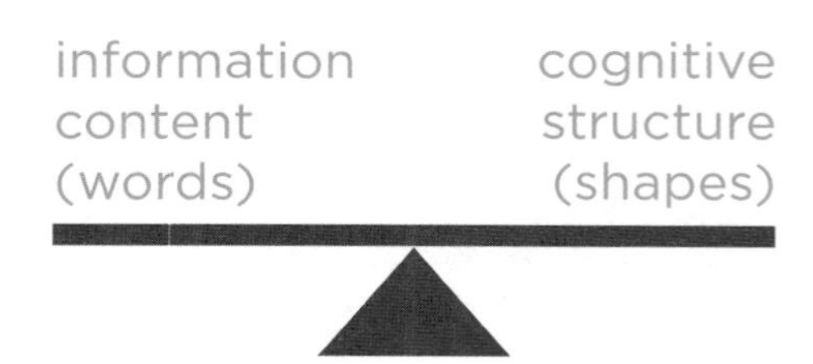

Figure 4.2: Metacognitive Maps Are Balanced

Metacognitive maps *balance* cognitive structure (shapes) *and* information content (words), unlike most diagramming and visualization techniques. Metamaps highlight the "universal cognitive grammar" that underlies our systemic thought and makes it more explicit, thereby leading to better systems thinkers. A metamap prioritizes balance between structure and information by giving ample real estate to the shapes that make up the underlying structure. In contrast, most maps prioritize the informational content and let text length dictate the size of the shapes.

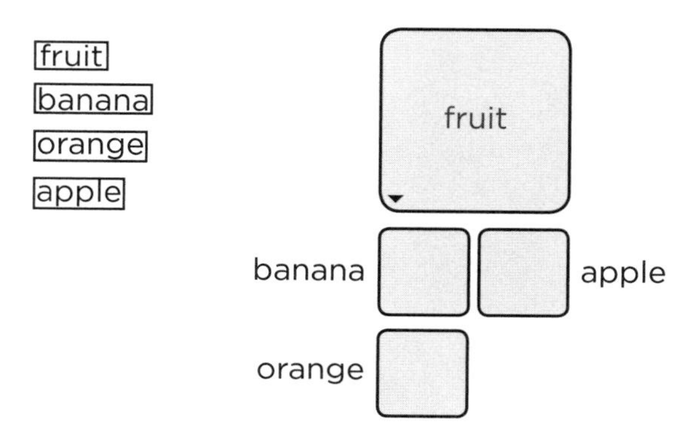

Figure 4.3: Traditional Vs. Metamap Sizing

REDUCE VISUAL AND COGNITIVE AMBIGUITY

Edward Tufte has shown that the cognitive style of PowerPoint software and the ambiguity it can create in the communication of information was a significant cause of the Columbia Shuttle disaster.[2] We share this because it otherwise might seem hyperbolic to claim that the cognitive style of a diagramming, mapping, modeling, or presentation method could have life and death consequences.

The DSRP rules that underlie metamaps increase the fidelity of mental models to reality. Additional thought went into designing the metamap diagramming method in order to decrease visual and cognitive ambiguities. This research focused on existing mapping methods.

Optimal ways to convey ideas

Simple visuals can make an enormous difference in understanding complex systems and solving wicked problems because human cognition (thinking) is *embodied*. The *Cortical Homunculus* (meaning "little cortex man") is a scientific model of the body in terms of the proportion of neurons allocated to various body parts. What one sees quite readily is the massive role that the hands and eyes play in both the sensory and motor cortex.

Cortex Man illustrates the importance of visualization (seeing with the eyes) and tactile manipulation (grasping with the hands). Our research lab developed visual and tactile modeling tools to serve the needs of systems thinkers.[1]

[1] Cabrera, D., & Colosi, L. (2010) "The World At Our Fingertips: The Sense of Touch Helps Children to Ground Abstract Ideas in Concrete Experiences." *Scientific American Mind* Volume 21, No. 4, September/October 2010, p36-41.

Cortex Man

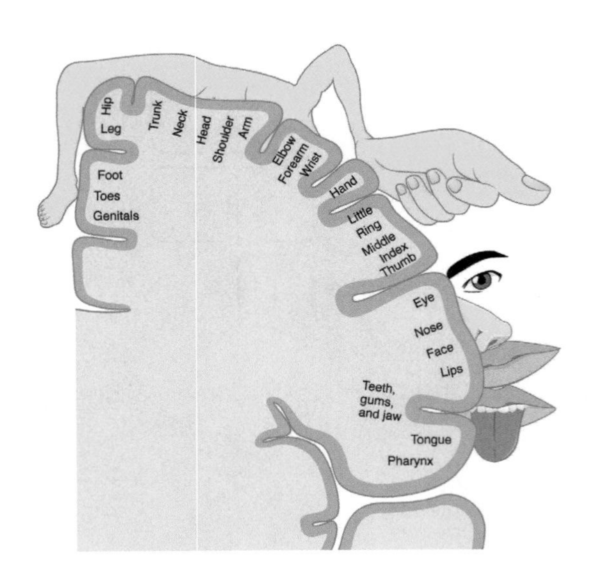

Cortex Man

Tools for systems thinkers should be both visual (MetaMaps) and tactile (ThinkBlocks). The simple rules of systems thinking have been programmed into 2D and 3D tools to help people learn how they think systemically about anything.

ThinkBlocks are 3D dry erasable, nested, relational, and perspectival blocks that are tactile so people can do systems thinking in groups, using their hands to move ideas around.

Systems thinkers use visualization and object manipulation not only because systems are complex and problems are wicked, but because the human mind is intrinsically programmed to understand through visualization and tactile manipulation.

ThinkBlocks

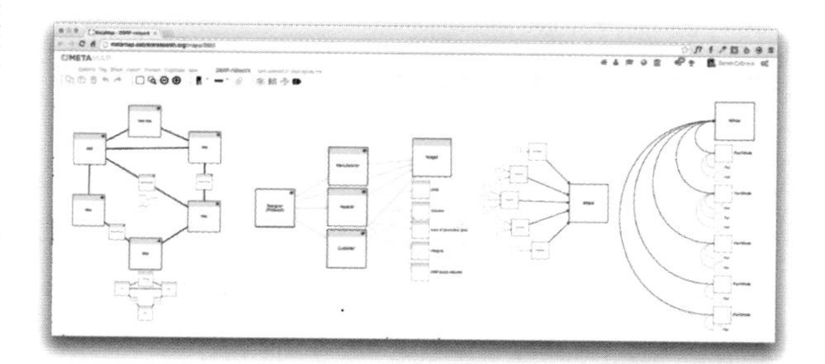

MetaMap Platform

ABUSE OF COLOR

Type "graphic organizer" or "systems thinking" into a Google image search and you'll see thousands of images of visual organizers people use to better understand or communicate their ideas. At first glance, they all look different, but if you look more closely for a structural pattern among them, you'll see they can be boiled down to a handful of *types* of maps such as hierarchical (used for organizational charts), radial, flow maps, tree structures, Venn diagrams, and feedback loops.

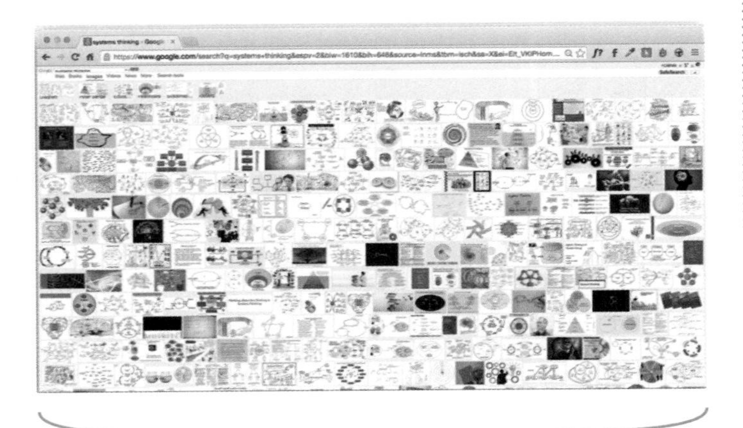

DSRP

distinctions systems relationships perspectives

Figure 4.4: DSRP Rules Underlie a Myriad of Maps

The four DSRP rules underlie all of these maps. We selected a small but diverse sample of the most popular systems thinking maps. Let's see how DSRP not only underlies them, but can help to make them even more robust and effective.

One of the things you'll notice is that color is used often. That makes sense because color is a powerful visual aid. Unfortunately, when we more closely inspect the use of color, we often realize that it is an abuse of color. For example, the popular style of visual mapping called mind maps often uses color, ostensibly to demonstrate creativity. Many who see the mind map in Figure 4.5 would conclude that it is creative, but few would look at the radial-tree diagram to the right of it and call it creative. Yet structurally the map on the left and the one on the right are exactly the same: a tree that radiates from a central point.

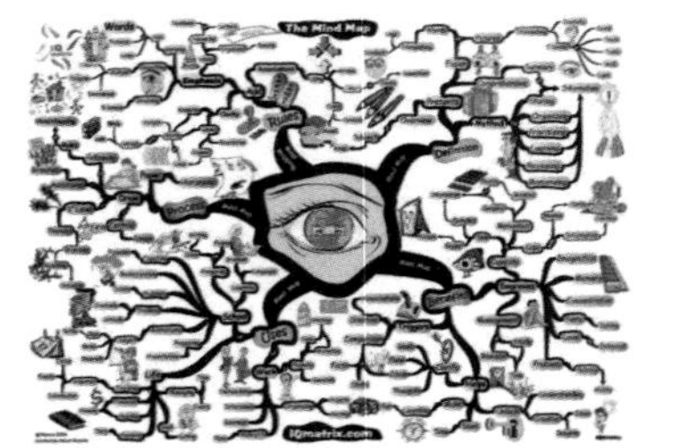

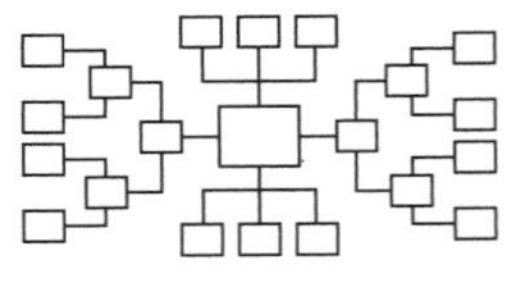

Figure 4.5: When Color Obscures

Notice too that the green, blue, red, and purple areas of the map simply throw color onto existing structure. The color detracts from rather than adds to clarity. In contrast, metamaps use color dynamically to indicate where the map can grow based on DSRP simple rules. Color is not abused but used judiciously for the most important structural aspects of thinking.

ABUSE OF SHAPES

Take a look at the structure of nearly any map and you'll see something similar about all of them. They use shapes, usually as placeholders for informational text or labels.

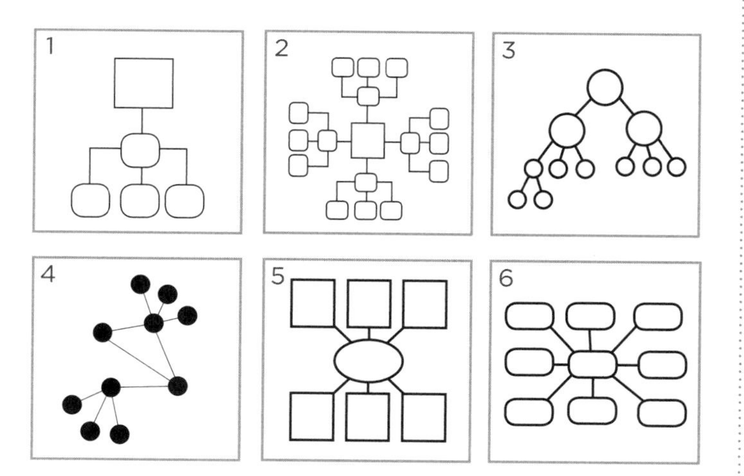

Figure 4.6: Traditional Visual Maps

When you look a little closer you'll realize that there's something these maps have in common: all the maps have *things* of some kind, usually represented by a shape or a label, or both. Some maps use the label itself as an object. These objects are distinguishing one idea from another. Visually speaking, it's as if the map uses the boundary of the object to say something about the thing: "I'm a distinct idea worthy of getting my own shape."

Figure 4.7: Multiple Shapes Used in Visual Maps

Is there an advantage to using one shape over another? Should we use the full palette of shapes or stick to one? At its core, there is no significant difference between using one shape over another: say a square versus a circle, for example. But there is an advantage to considering all shapes not as the thing they represent but as representations themselves. That is, every shape in every conceptual map ever produced is a concept, not the thing it represents. This idea is best explained by the Belgian painter René Magritte, in his painting of a pipe under which he wrote (in French), "This is not a pipe." He remarked about the painting:

> The famous pipe. How people reproached me for it! And yet, could you stuff my pipe? No, it's just a representation, is it not? So if I had written on my picture "This is a pipe", I'd have been lying![4]

[4] Magritte, R., & Torczyner, H. (1977). *Magritte, ideas and images* (p. 71). New York: H.N. Abrams.

Figure 4.8: Representation

Likewise, the ideas or concepts we put in a systems thinking diagram are not the actual things they represent, but are representational. It is extremely important to remember that these concepts are mental models, representations of real things, not the real things themselves. If everything in a systems thinking map is the same thing—a representation—then it should be denoted by the same shape. Which shape we choose is irrelevant, but the square or circle is a good shape for several reasons. The triangle (and to a lesser extent the circle) does not take longer labels as easily (Figure 4.9), leading to aesthetically ugly maps. The rectangle is a good choice because it makes use of the space required for the label but for the very same reason it is a poor choice. In practice, when used in systems diagrams, rectangles are manipulated to create a number of different sized shapes based solely on the arbitrary textual length of the label. Therefore, we lose one of the most important visual cues we have— size—to an arbitrary feature (text length).

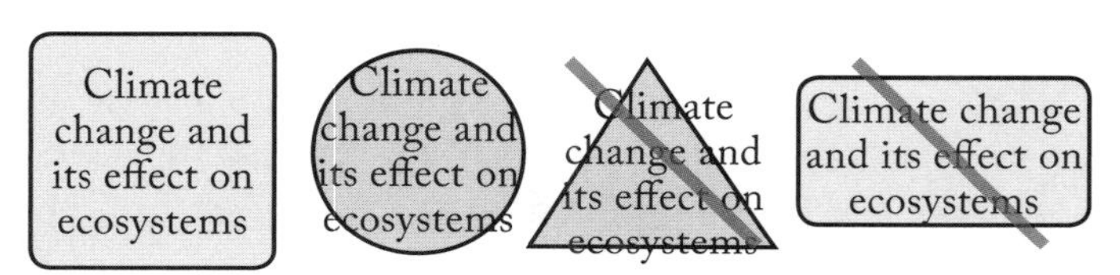

Figure 4.9: Shape Matters

Choosing a single shape to represent concepts is a big step in developing a universal visual grammar that communicates using similar structures across disparate topics and systems. If everything in a map is a distinction, then why not distinguish it with what actually differentiates it: its information content. In this way, we can understand something important about the structure of all systems mapping: that every object in every map is a distinction and that these boundaries must be questioned.

ABUSE OF LINES

To use a line to connect two shapes is human and to think of that line as a relationship is natural. But you will notice in many figures that lines are used to communicate many things and sometimes nothing at all. Lines should be afforded exclusive status as indicating relationships.

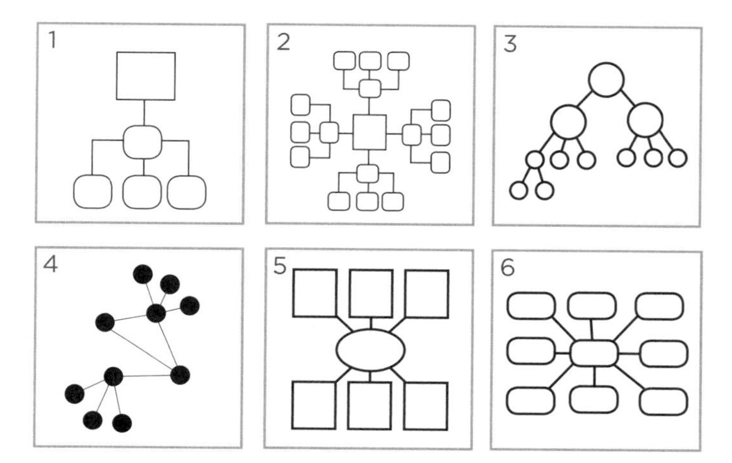

Figure 4.10: Line Confusion

More often than not, visual diagramming techniques use lines for part-whole relationships rather than the many other types of relationships that might occur between two ideas or things. This leads to a lot of visual ambiguity and confusion. In Figure 4.10, you can see six extremely popular styles of maps. All are hierarchical tree diagrams or network diagrams that more often than not represent part-whole containment but can also represent real relationships because the lines are being used ambiguously. For example, the lines in map (1) could represent the direct reporting relationship of an employee to a boss or they could represent parts of a whole. By far the most popular visual abominations, Mind Maps (shown in 2, 5, and 6), misuse lines as part of their basic architecture, resulting in ambiguous maps. Ask yourself, in the six visual maps in Figure 4.10, do lines represent part-whole structure or relationships between the objects?

ABUSE OF SIZE

Size should matter, but it often doesn't. Notice that the variety of maps in Figure 4.10 may—or may not—illustrate a hierarchical structure (perhaps the most popular type of structure). All hierarchical structures are part-whole in nature. Part-whole structures allow us to represent systems nested within systems, or *subsystems*. Using Venn diagrams we can see the basic idea of part whole structure is that the part belongs to or is contained by the whole as in Figure 4.11.

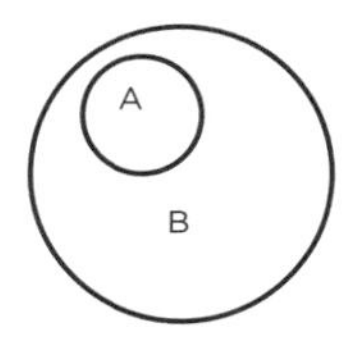

Figure 4.11: Containment

The beauty of Venn diagrams is that they make elegant sense. In the case of the example in Figure 4.11, it makes sense to us that A is inside of B because A is smaller than B and B contains A. The reason for this is that size is used. Our mind is accustomed to the parts of things being smaller than the things they are a part of. Therefore, even when this part-whole structure is metaphorical as in the case of abstract ideas, the mind readily understands it. Our tendency to abuse lines and to abuse size are related and can be resolved. Reserve lines exclusively for relationships. That way, when you see a line, you'll assume there is some relationship to be considered between two ideas, rather than being confused as to whether it's a true relationship or part-whole structure. And reserve shape-size exclusively for part-whole containment.

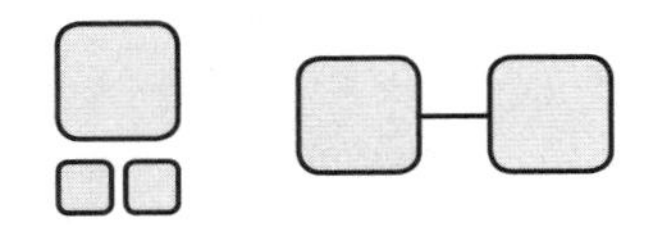

Figure 4.12: Line and Shape-Size Clarity

For example, in Figure 4.13 we reproduce the visual ambiguities of map 3 in Figure 4.10. It is impossible to know exactly what the author of such a map intends to communicate. The size differences of the original seem to indicate that there may be some part-whole structure but the top three circles are the same size. The map could be interpreted as a straightforward hierarchy or as a relational network.

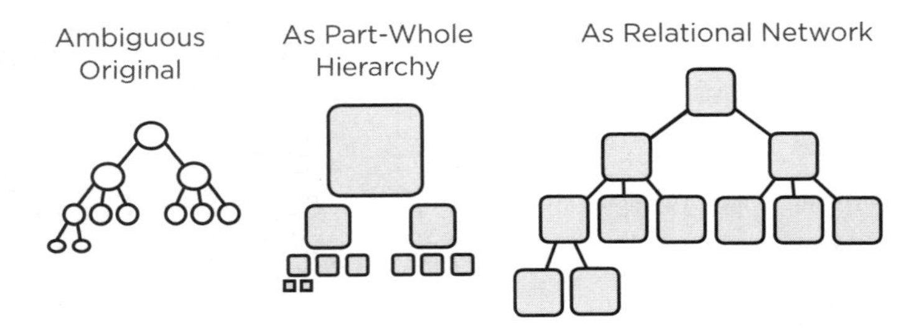

Figure 4.13: Use Size for Part-Whole, Lines for Relationships

Developing a more universal visual grammar will increase understanding and aid communication. Let's review the advantages we are suggesting:

1. use the same shapes for all distinctions;
2. use lines exclusively for relationships; and
3. use shape-size exclusively for part-whole (see Figure 4.14).

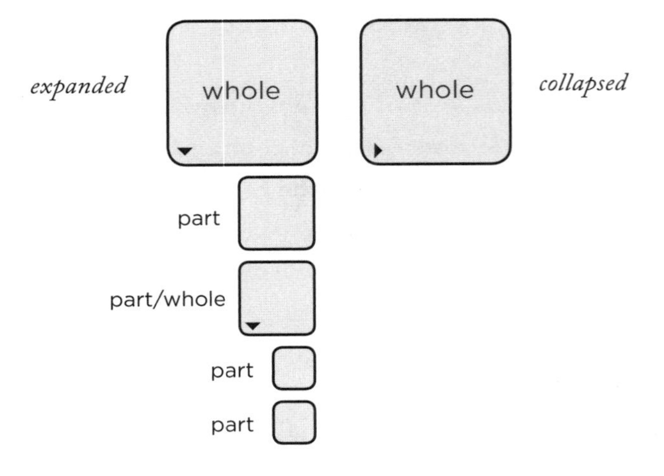

Figure 4.14: Size Matters

A triangle in the lower corner of a square (Figure 4.14) indicates that a concept has parts and whether those parts are expanded or collapsed for viewing purposes. This expand/collapse view is an important cognitive feature that allows the mind to perform "chunking"[5] operations, which increase comprehension and the degree of complexity the mind is capable of handling.

Using relative size for part-whole structure rather than lines makes good sense, and it also frees up lines for what they were designed to do—visually relate things.

USING LAYOUT

The spatial layout of parts can communicate several things. In Figure 4.15 the three maps are the same except in that they use different layouts. In the left map, the parts are laid out in a "freehand" network of relationships. In the center, an "outline" view shows relations between the parts, which could be justified left or right. On the right, a linear progression of related parts indicates steps, or a series in time.

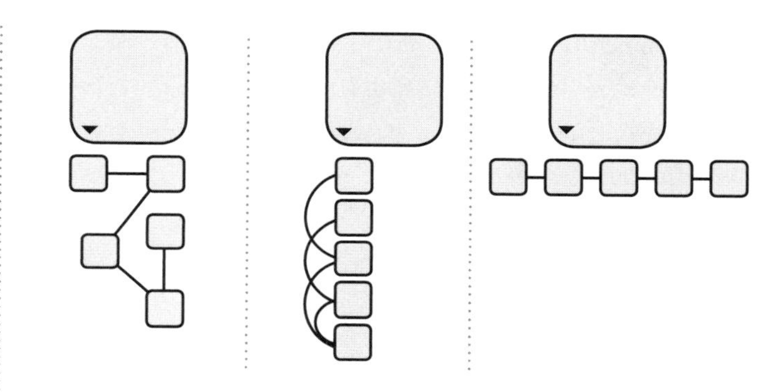

Figure 4.15: Layout Options

MIX AND MATCH SIMPLE RULES

DSRP is universal to all systems thinking and metamaps offer a simple but effective palette (Figure 4.16) for visual clarity using the principles above. The adaptive nature of metamaps is based on the user's ability to mix and match and combine and recombine these basic visual elements.

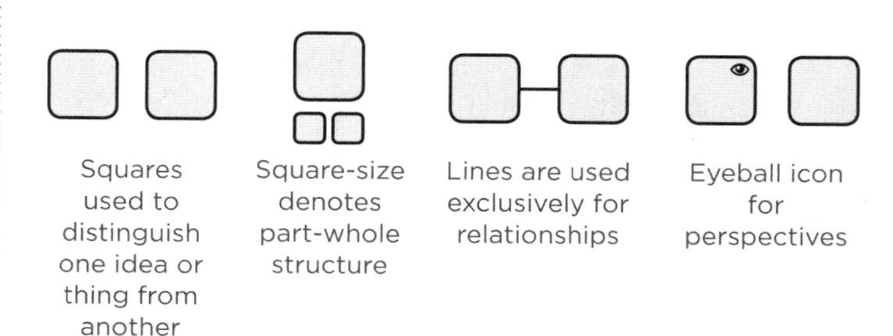

Figure 4.16: MetaMaps Palette

[5] Miller, G. A. (1956). "The magical number seven, plus or minus two: Some limits on our capacity for processing information". *Psychological Review* 63 (2): 81–97. doi:10.1037/h0043158.; Simon, H.A. (1974) How Big Is a Chunk?: By combining data from several experiments, a basic human memory unit can be identified and measured. *Science*. Feb 8;183(4124):482-8.

Now imagine that, like Lego bricks, these 4 simple DSRP rules can be combined and recombined in infinite ways. For example, in Figure 4.17, the distinctions rule and the systems rule could be combined to make a new thing that is also part of an existing part-whole system. You might imagine that this new map represents a part of a spark plug, which is part of the engine, which is part of the car.

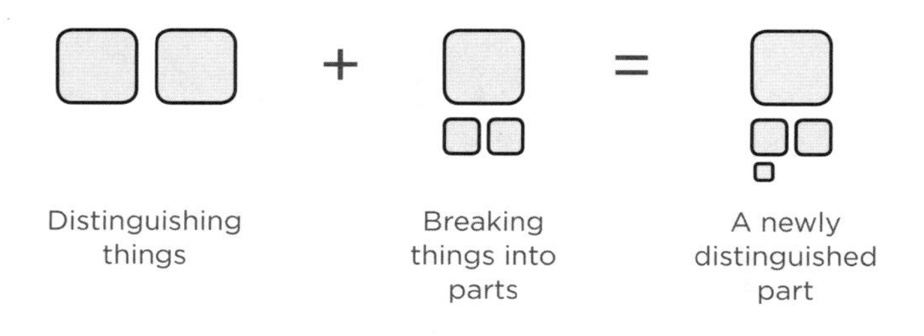

Figure 4.17: Combine Distinguishing Things with Breaking Wholes Down into Parts.

There's no limit to what you're allowed to do with these four DSRP rules and their visual representations. As another example (Figure 4.18), we could take the "whole with two parts, one of which has a part" and combine it with a relationship to get a distinctive and systematized relationship between two things. So in this case, our end product might be the relationship between biology and chemistry, which makes up biochemistry—a new discipline, itself made up of many parts. Or, alternatively, you could think of the relationship as the multifaceted conflict between Palestine and Israel.

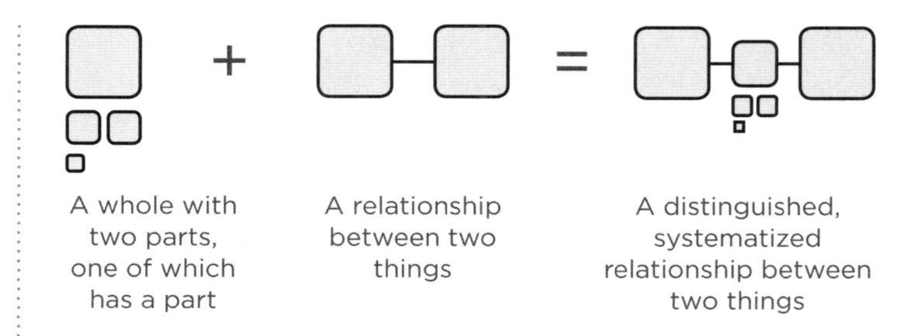

Figure 4.18: Add a Part-Whole System onto a Relationship

And any one of the objects in any map can be the point or view of a perspective, which can be mixed and matched. Figure 4.19 illustrates how the result of our last combination can be combined with the perspective rule to create something new. In this case, we could think of this map as representing the very different perspectives of Palestine and Israel on the complex, multifaceted relationship between them.

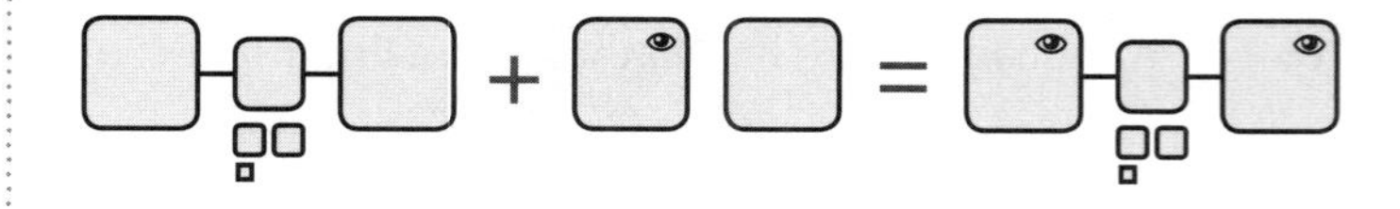

Figure 4.19: Combine the Output of 4.18 with Perspective

Let's do one more. Apply the relationship rule to the parts of a whole and you can get an ecology of interrelated parts. This could represent the related parts of a social group, a truck, or an actual ecosystem—anything that has a bunch of interrelated parts.

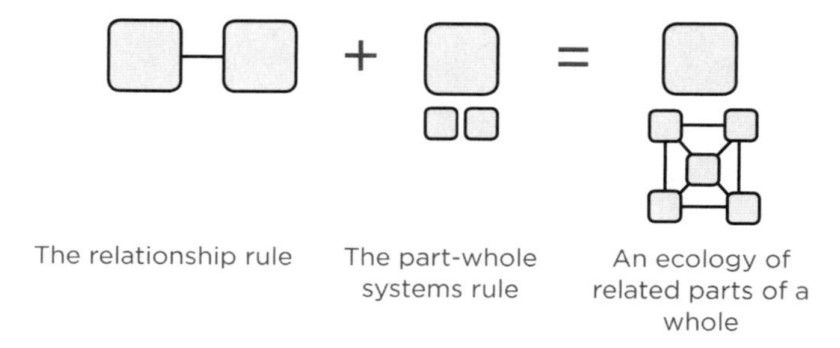

Figure 4.20: Apply Relationship Rule to the Parts of a Whole

And so systems thinking requires little more than practice in building with these cognitive building blocks. It is really no different than building with four different types of Lego, or four different nucleotides, or four base colors. Mix and match. Combine and recombine. Have fun. There is an infinite number of thoughts you can think and systems to explore, but these simple recombinant rules and the visuals we have outlined here make it easy to visually represent any system, no matter how complex it might be.

CHAPTER 5

USE AND REUSE COGNITIVE JIGS

COGNITIVE JIGS

A cognitive jig is a common underlying structure of systemic thought. The benefit of cognitive jigs is they can be used over and over again to create meaning and understanding. Also, they are content agnostic, which means they can be used across any topic, domain, or discipline. Use of cognitive jigs saves time, increases the speed of thought and understanding, and facilitates interdisciplinarity. Whereas DSRP structures are universal (ever-present in our thinking), cognitive jigs are seen frequently, but not always. We can utilize them when they are useful.

Cognitive jigs are content agnostic, which means they can be used across any topic, domain, or discipline.

You've used cognitive jigs countless times before—perhaps without being aware of it. There are four that you're already quite familiar with: analogies, metaphors, similes, and categories. DSRP shows us that there are many more of these cognitive jigs, each one as useful as an analogy, which humans use a lot to better understand things.

We've given these new cognitive jigs (or jigs for short) descriptive and sometimes humorous names, but they serve a serious purpose. Imagine a time in the future when the young people of today—as grown adults—purposefully use jigs they casually refer to as "P-circles" and "R-channels" and "Part-parties" as often, and with as much benefit for communication and understanding, as we use metaphors and analogies today.

First, we will review a few common jigs you already know to see their fundamental DSRP structure. These include:

- Analogies
- Metaphors
- Similes
- Categories

Then, we'll introduce a few more common structures you will find useful:

- P-Circles
- Part-Parties
- Barbells
- R-Channels

Imagine a time in the future when the young people of today—as grown adults—purposefully use jigs they casually refer to as "P-circles" and "R-channels" and "Part-parties" as often, and with as much benefit for communication and understanding, as we use metaphors and analogies today.

Jigs

Jewelers, carpenters, welders, metalworkers, architects, designers, attorneys, cave men, web programmers, and quilters use jigs, so why not systems thinkers? Jigs go by other names: stencils, patterns, templates, overlay, boilerplate, and guides.

A jig's primary purpose is to provide replicability and accuracy in the creation of products. Jigs are of immense utility to carpenters and builders of all kinds. Jigs are used whenever something has to be produced again and again so that the creator doesn't need to reinvent the wheel every time. A jig is, in one sense, the beginning of automation.

Having used jigs in carpentry to automate the workload, Derek wondered whether there might be "cognitive jigs"—a kind of reusable structure that one might see again and again across different problem sets and topics. It turns out that there are many cognitive jigs and that they are immensely useful to the systems thinker.

You've likely used jigs before if you've used stencils or templates. You know those green plastic sheet stencils you get at office supply stores that make it easy for you to draw common shapes? That's a type of jig. In this chapter, we are talking about cognitive jigs that save us cognitive effort and increase the speed of thought.

Carpenter's Jig

Hand as Stencil

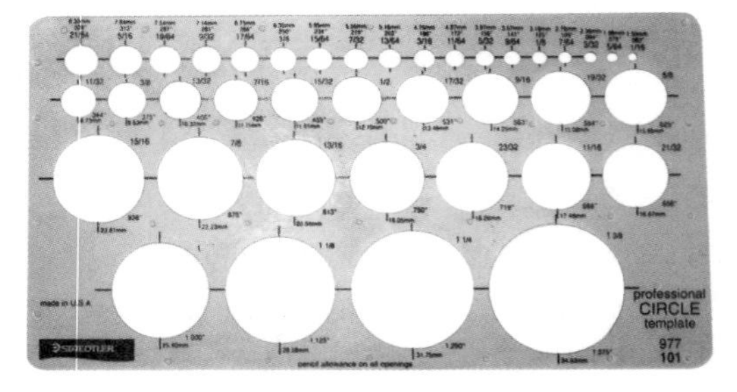

Stencil

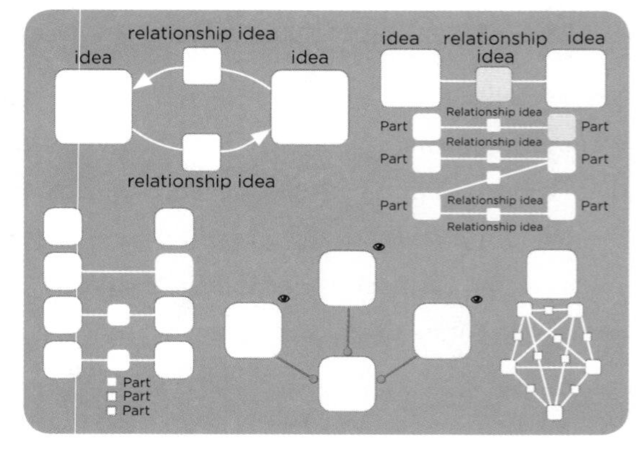

Cognitive Jigs

ANALOGIES, METAPHORS, AND SIMILES

We can't remember what it was like before the invention of the analogy. Was there a time before analogies?

Figure 5.1: The Very First Analogy?

Analogies are like iPhones. They are a form of technology—a conceptual tool that was invented to increase our ability to communicate and to understand.

Stop. Did you notice anything? We just made an analogy, but I suspect that your brain processed it like your stomach digests chicken soup—easy. (Oops, there's another one.) We can hardly refrain from using analogies because they are, like oxygen, pervasive. Here's the analogy I made above in a visual form:

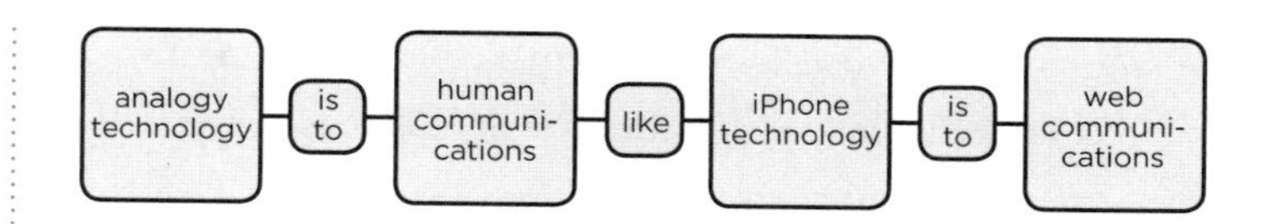

Figure 5.2: DSRP Structure of an Analogy

To fully grasp the important technological implications of the iPhone or Facebook or the Internet, we need to be able to compare and contrast before the invention and afterward. That's why it is so difficult to grasp the important role that cognitive jigs like analogies play in our lives: because none of us were alive before analogies existed. We can remember before email, but we can't remember before analogies. Yet in comparison, the impact of the analogy dwarfs the iPhone or Facebook. In 2013, over 300 million iPhones were in active use. Facebook had 1.4 billion active users in 2014. In contrast, there are over 7 billion people in the world, all active analogy users.

What makes the analogy such a remarkable innovation? There are a few things about analogies that make them advanced technology:

Analogies are cognitive jigs. The cornerstone of analogy-technology is that it is a common cognitive structure. The genius behind the invention of analogies was that they gave us a mental model of a common way we understand things (i.e., by comparison to a known thing).

Analogies are content agnostic. The remarkable contribution of analogy-technology is that it makes no difference what the content of the analogy is, the underlying structure of the analogy is generalizable and abstract. The nature of the relationships in an analogy remains the same for all analogies (namely, A *is to* B *as/like* C *is to* D). For example: doctor is to hospital as lawyer is to courtroom and son is to dad as daughter is to mom are both analogies and share the same cognitive structure. In Figure 5.3, notice that all of the structure stays the same and that only some of the content differs. For example, the structure (in orange) is the same for all three analogies, whereas the content varies in some places (black text) and remains the same in others (orange text). This is a pattern you will see in all cognitive jigs: either the content can change or the structure (or both or neither), but that which stays the same (e.g., the orange items) is the essential aspect of the cognitive jig.

Two other cognitive jigs (and subtypes of analogies) you are familiar with are similes and metaphors. As you can see in Figure 5.4, analogies, similes, and metaphors share common information and structure (in orange) and also have some differences (black).

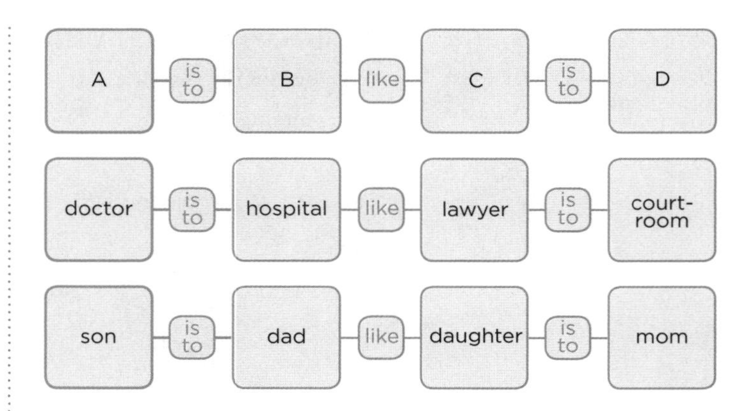

Figure 5.3: Variants and Invariants in Content and Structure

Figure 5.5 shows how these jigs work with the help of Shakespeare. Shakespeare said in *As You Like It*, "All the world's a stage, and all the men and women merely players; they have their exits and their entrances."[1]

Shakespeare was making a metaphor that the world is a stage. He was not making a simile: the world *is like* a stage. Inside his metaphor, he was also implying an analogy between the relationships between men and women and deaths and births and players (actors) and their exits and entrances.

[1] Shakespeare, W. (1954). *As you like it*; ([Rev. ed.). New Haven: Yale University Press.

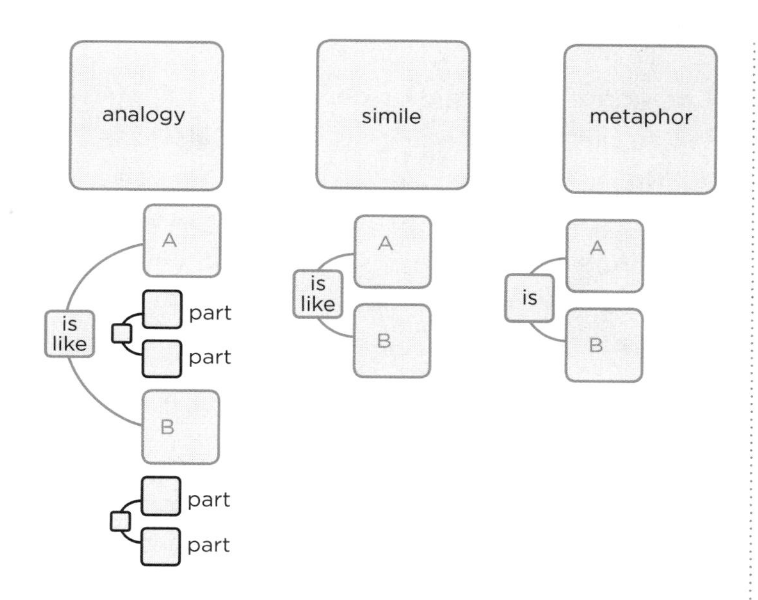

Figure 5.4: Commonalities and Differences in Content and Structure

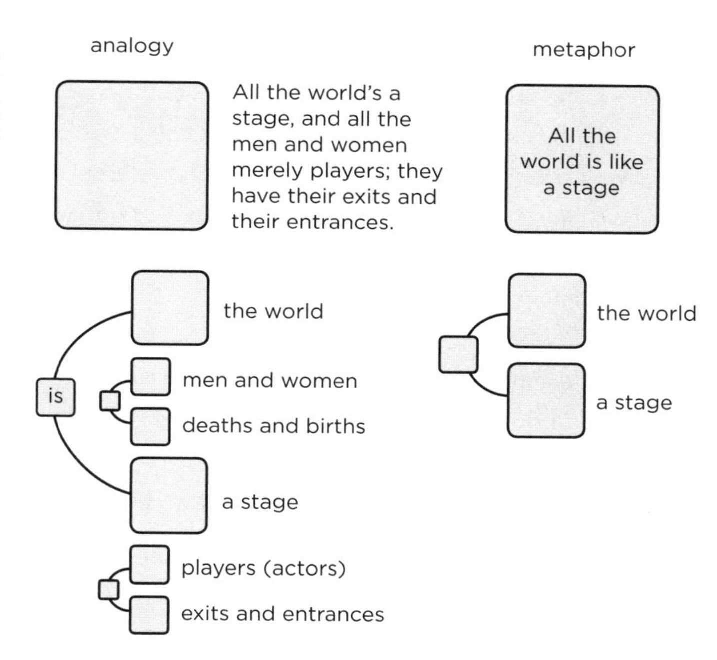

Figure 5.5: Mapping Shakespeare

We can see from these examples (Figures 5.2 through 5.5) that DSRP patterns underlie the common structures of cognitive jigs, at least in the case of analogies, similes, and metaphors. But what about another very popular cognitive jig, categories?

CATEGORIES

Analogies, similes, and metaphors are powerful cognitive jigs in large part because they help us to relate unfamiliar things with familiar things. But there's another cognitive jig that you are already familiar with: categories. Categories have a similarly long history to analogies, but categories have become insidious.

We could say similarly nice things about the marvelous "invention" of the category as we did about analogies. The invention of the category dates back to Aristotle (B.C. 384–322). Today we can hardly imagine what it would be like to exist in a world unstructured by categorization. Like analogies, categories have much to be admired. It is common parlance in the field of cognitive science to say that categories are universal cognitive structures—that we cannot survive nor have a thought without them. Yet categories have a dark side: they are a cognitive *cul de sac* or dead end in the road. Categorization makes us feel like we are getting somewhere, speeding down the highway of understanding and knowledge until *Wham!*, dead end. We're stuck. And, it usually takes a long time to get unstuck.

“Are you serious?” That's usually the reaction we get when we say that we are suspicious of categories. After all, categorization has given us so many gifts of understanding:

Categories have a dark side: they are a cognitive cul de sac or dead end in the road. Categorization makes us feel like we are getting somewhere, speeding down the highway of understanding and knowledge until *Wham!*, dead end. We're stuck. And, it usually takes a long time to get unstuck.

- The scientific disciplines (e.g., physics, chemistry, biology, psychology, sociology, economics);
- The hierarchy of biological classifications (Life, Domain, Kingdom, Phylum, Class, Order, Family, Genus, Species);
- Bloom's Taxonomy (Knowledge, Comprehension, Application, Analysis, Synthesis, Evaluation); and
- The organization of the Internet, your computer desktop, office filing systems, libraries (directory structure, file folders, Dewey decimal system, etc.)

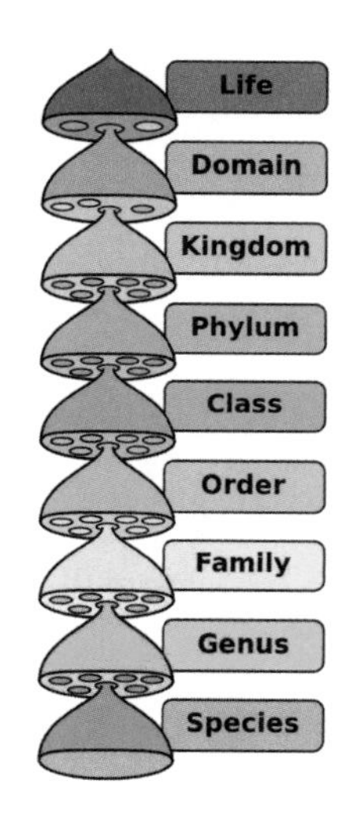

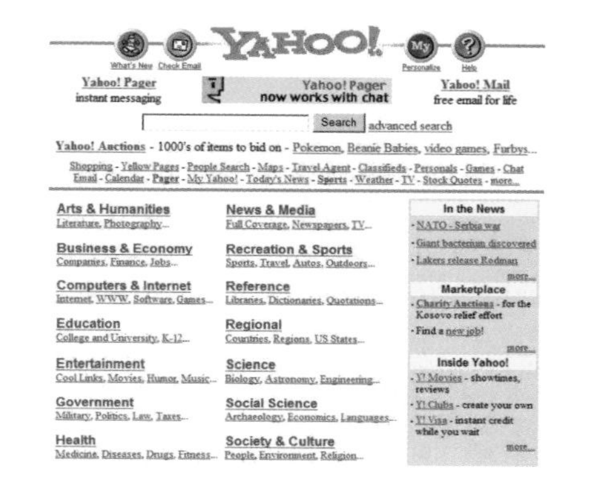

It is estimated that there are up to 100 million species. If you drew one triangle every second to represent a single species, it would take you 3.17 years to finish. To organize this massive collection, scientists tried categorization, but it failed. Instead, they group organisms into species according to 21 scientifically valid perspectives. How do scientists know which perspective to use? It depends on what they are trying to do. Part-whole groupings based on different perspectives are universally accepted; categorization is not.

When the web was first organized, it was categorized using vast directories like the folders on your desktop or books in a library. Searching seldom resulted in finding. Google revolutionized the web by thinking about it in terms of relationships between pages rather than discrete categories.

Figure 5.6: Outdated Categories

Categories make us feel like we understand the universe because we are able to cognitively capture it. The question is, do categorical structures adequately represent the real structure of the universe? Do categories help us feel or *be* more knowledgeable? Let's take another look at the above examples.

- ***The scientific disciplines:*** the categorization of knowledge into scientific disciplines did a lot to propel scholarship and knowledge accumulation, but today it poses us real problems. This is because the universe we are trying to understand and the problems we are trying to solve do not heed our disciplinary boundaries. The universe and all of the problems in it are interdisciplinary. University policies and financial structures, departmental culture and tenure structure, and our own thinking is hindered by the categories we set up. Our categorization of knowledge into disciplines has turned into a *cul de sac* that is going to take a long time to escape.

- ***The hierarchy of biological classifications:*** we teach it in schools as if it's fact but the "species concept" is wildly complex and nothing like the streamlined categories we all learned in grade school. Today, scientists have not 1 but 21 equally valid and useful perspectives on how to group organisms into species, including morphologically, ecologically, genetically, and through mate recognition. Which perspective the scientist uses depends on the job at hand. Bacteria pose significant problems for categorization. What used to be a biological freeway to understanding has become a categorical *cul de sac*.

• ***Bloom's Taxonomy:*** very few ideas are as popular or influential in education today as Bloom's. Yet few teachers can put it to practice and research validates none of it. Despite this, it was recently revised, demonstrating our allegiance to categories despite their infidelity to reality. Knowledge and higher-order thinking skills are far too robust and complex to stuff into Bloom's discrete categorical prison. It's going to take years to retrain teachers to get out of this *cul de sac*.

• ***The initial organization of the Internet, your computer desktop, office filing systems, the Dewey decimal system:*** filing systems, computer desktops and even the world wide web have been organized using categories dating back to their origins. That's exactly why no one can find anything. Another categorical *cul de sac*. Take for example the early organization of the Internet, which directories like Yahoo! organized by categories. Searching seldom led to finding. Google revolutionized our ability to find things through keyword searches and relational networks. Likewise, searching by keyword for files on your computer is far more successful than searching through nested folder systems only to realize that the mindset you're in at present isn't the same mindset you were in when you saved the file to a particular folder.

Worst among the impacts of blind category use are the categorization skills we teach to children that lead them to be less robust, more black-and-white, and less adaptive thinkers. We say that categories are *insidious* because it is a fitting description of how they operate: "proceeding in a gradual, subtle way, but with harmful effects."

What DSRP structure reveals is that the application of discrete categories to real-world phenomena is inadequate to fully understand something. Instead, we must see that all categorization is based on a perspective (which is usually not made explicit). If we are to escape the category *cul de sac* but still benefit from its use we should replace static categories with part-whole systems grouped dynamically by perspectives.

NEW JIGS

With the remarkable impact of such common jigs as analogies, metaphors, and similes, one wonders why haven't more been created? DSRP reveals more common structures and we will review a brief list of some of them. Many will seem familiar because you're already using them, but like analogies and categories, *when we are conscious of these common jigs, they become even more powerful as they open up new avenues of understanding and communication.* The next section introduces several important cognitive jigs that can be used and reused to gain deeper understanding of various topics, issues, and problems. These jigs are, like analogies, agnostic to content. We will cover P-circles, Part-Parties, Barbells, and R-Channels.

P-CIRCLES

Remember that perspectives in metamaps are shown by an eyeball on the top right corner of any square. This means that an idea or thing is "viewing" some other thing or idea. Figure 5.7 reads, "an idea (b) from the perspective of idea (a)." Remember also that a perspective is comprised of a point (here, b) and a view (here, a).

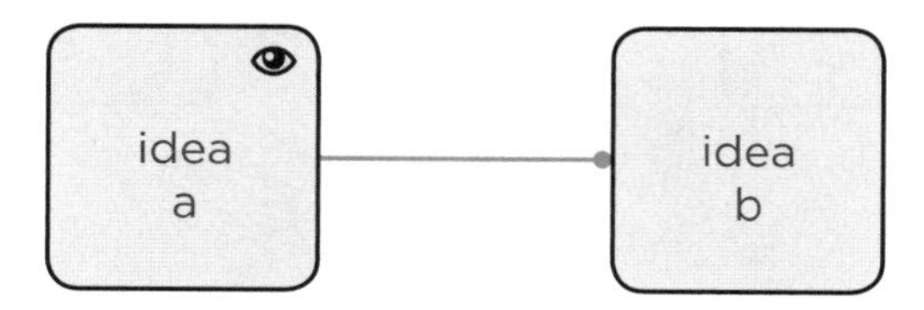

Figure 5.7: Perspective Diagram With Point and View

Changing the point or the view changes the perspective. Most of our thinking errors come from thinking that the point is the same as the perspective when it's not. A perspective is made up of both a point and a view. So, while there are only two things in Figure 5.8 (Dad and Me), they can be configured into four different perspectives depending on which thing occupies which position (point or view) of the perspective. C and D show dad's perspective on himself and my perspective on myself. In these cases, the point and the view are the same (i.e., self-reflection).

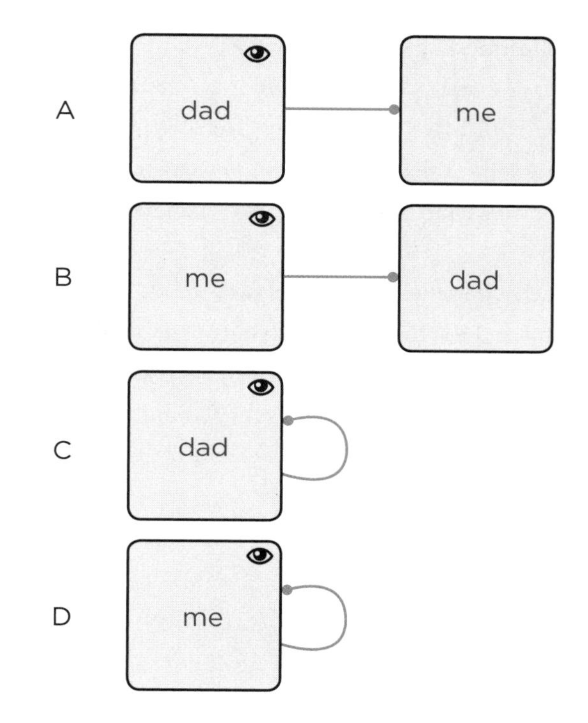

Figure 5.8: Four Perspectives

If we are to escape the category *cul de sac* but still benefit from its use we should replace static categories with part-whole systems grouped dynamically by perspectives.

Figure 5.9 illustrates how combining a view with multiple points can create multiple perspectives in a cognitive jig we call "Perspective Circles" or P-Circles. The idea is that there is some thing or idea that is the view and we take a look at that thing or idea from multiple points simultaneously.

We represent taking multiple perspectives by adding eyeballs on each point and also distinguishing between them (i.e., different squares). The P-Circle in Figure 5.9 reads, "an idea (a) viewed from point (b) and point (c)." A simple example would be looking at a new product (a) from the point of view of marketing (b) and engineering (c).

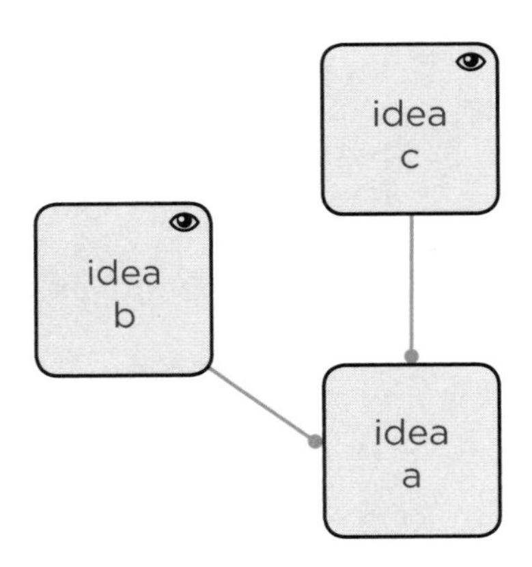

Figure 5.9: P-Circle with 2 Perspectives

Figure 5.10 shows another P-Circle which reads, "world hunger from three points of view: environmental, economic, and political." Because these cognitive jigs are content-agnostic, we could add any content of the form, "idea (a) from three different points of view: (b)(c) and (d)."

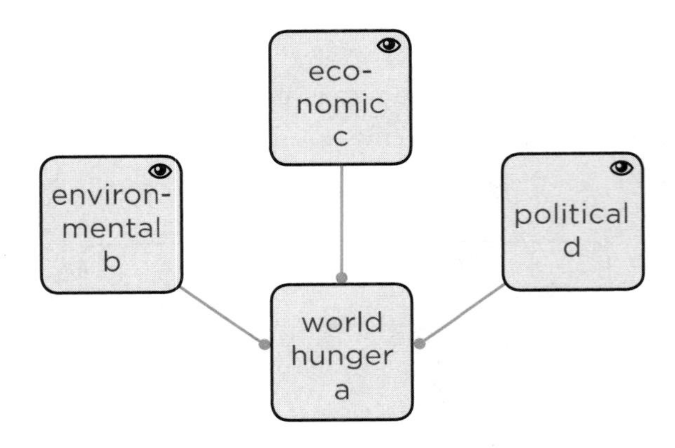

Figure 5.10: 3 Points of View on World Hunger

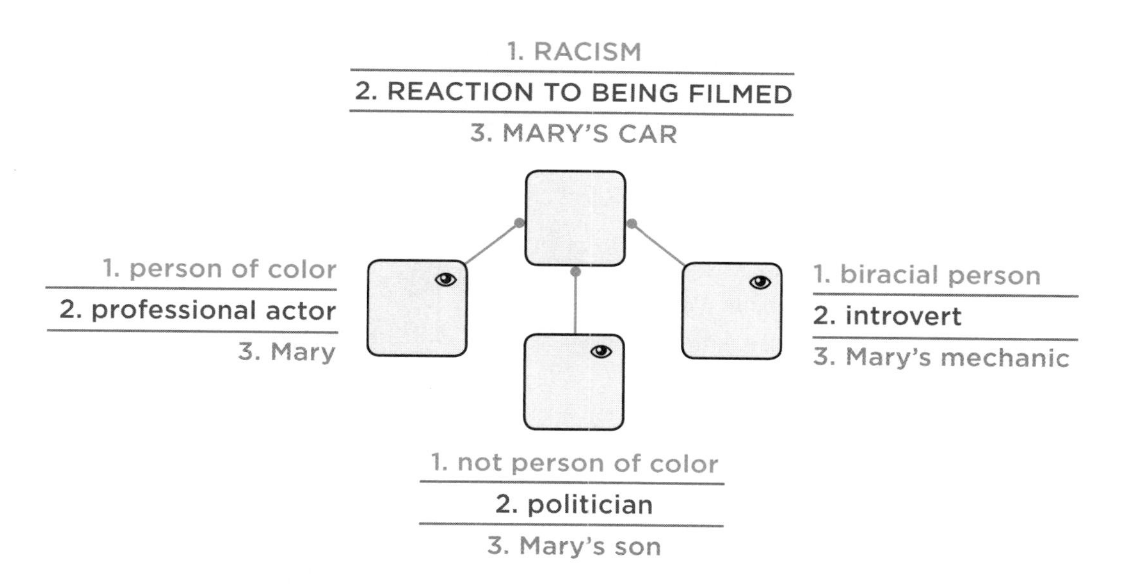

Figure 5.11: Three Examples of a Perspective Circle, Where One Thing is Viewed from Three Points of View

Note in Figure 5.11 that the same P-circle structure can be used and reused in many topical areas. The metamap shows a simple P-circle with three points on a single view. Because jigs are content agnostic, the same P-circle structure can be used across 3 wide-ranging examples which are distinguished by the color of their text (1, 2, and 3). For example, in the #1 orange example Racism is seen from the perspective of a person of color, not a person of color, and a biracial person.

A P-Circle is a remarkable jig with broad application. Any time we explore any idea we can take multiple perspectives on it.

Looking at something from different perspectives often yields different parts. The metamap in Figure 5.12 reads, "a whole idea (a) from three points of view (b)(c)(d), each yielding different parts (b1, b2, b3, c1, c2, c3, d1, d2, d3)." For example, a wicked problem, the AIDS epidemic, requires taking multiple perspectives to make headway on a solution. When we look at the AIDS epidemic (a) from the perspectives of epidemiology (b), government policy (c), and socio-cultural factors (d), you'll see many sub-issues that must be considered to reduce the worldwide incidence of AIDS. These sub-issues are brought to light by the different perspectives. For example, when looking at the AIDS epidemic (a) from the perspective of epidemiology (b) we know that we must attend to control of the infection (b1), patterns of spread (b2), and patient monitoring and tracking (b3). The government policy perspective (c) highlights the need for coordination among the Center for Disease Control (c1), World Health Organization (c2), and tribal governments (c3). Factors that need to be considered from a socio-cultural perspective include sexual practices (d1), group stigma (d2), and religious norms (d3).

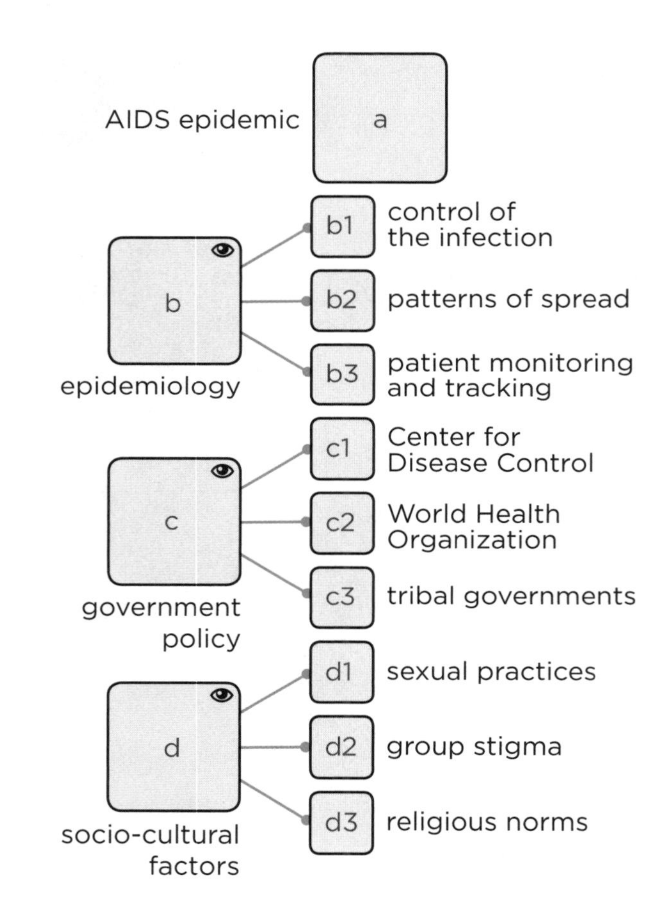

Figure 5.12: Parts of the View Seen from 3 Different Points (AIDS Epidemic)

Note in Figure 5.13 that the jig's basic structure is content agnostic. Three different content areas utilize the same basic structure.

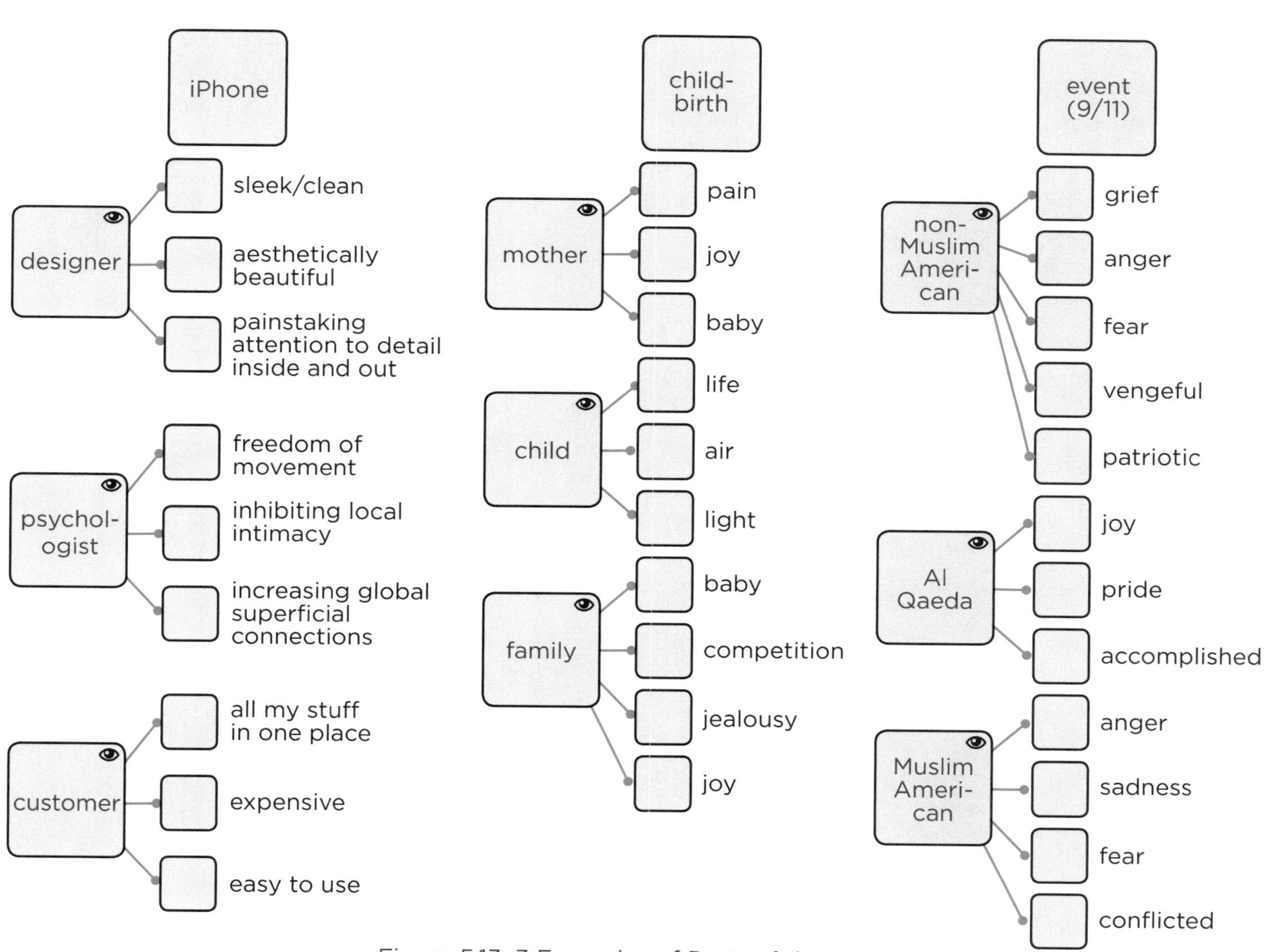

Figure 5.13: 3 Examples of Parts of the View Seen from 3 Different Points

SUB-PERSPECTIVES

It's important to note that the point of a perspective is not always homogeneous. In Figures 5.12 and 5.13 the points cause the view to have parts that are uniquely associated with a given point. In contrast, Figure 5.14 illustrates that the point can also be a whole made up of parts. The point (a) and the view (b) can contain any level of complexity. For example, if you are developing a new solution to a wicked problem like world hunger (b), then obviously figuring out a solution requires looking at the problem from many different perspectives (a), each of which might have subparts.

Figure 5.14: Points are Wholes With Parts

As another example, Figure 5.15 shows that we might think about the Muslim perspective on the War in Afghanistan.

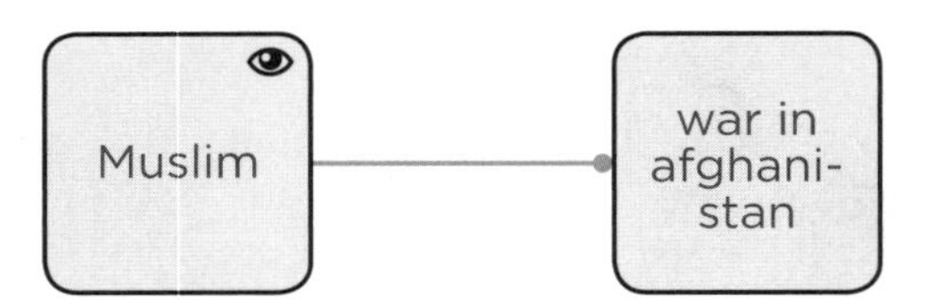

Figure 5.15: Homogeneous Perspective

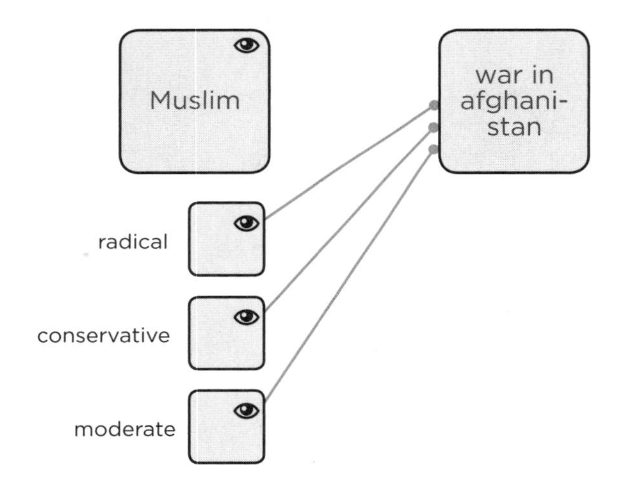

Figure 5.16: Heterogeneous Perspective

Thinking systemically in Figure 5.16, however, allows us to see that there is not a singular Muslim perspective on the war, but rather a number of sub-perspectives such as radical, conservative, and moderate Muslims, and in reality many more. The map in 5.16 shows this combination of perspective and part-whole systems. Solutions to messy problems require finer-grain analysis, including breaking the point of the perspective down into parts.

In Figure 5.17, we can see a single jig's basic structure remains the same but can be used across 5 different topical areas.

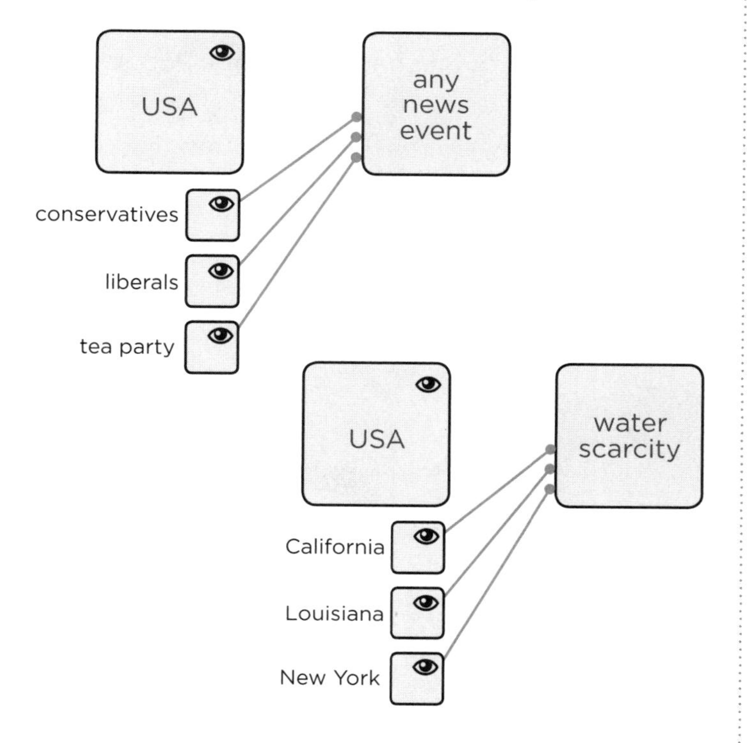

Figure 5.17: Agnostic Structure of Jigs

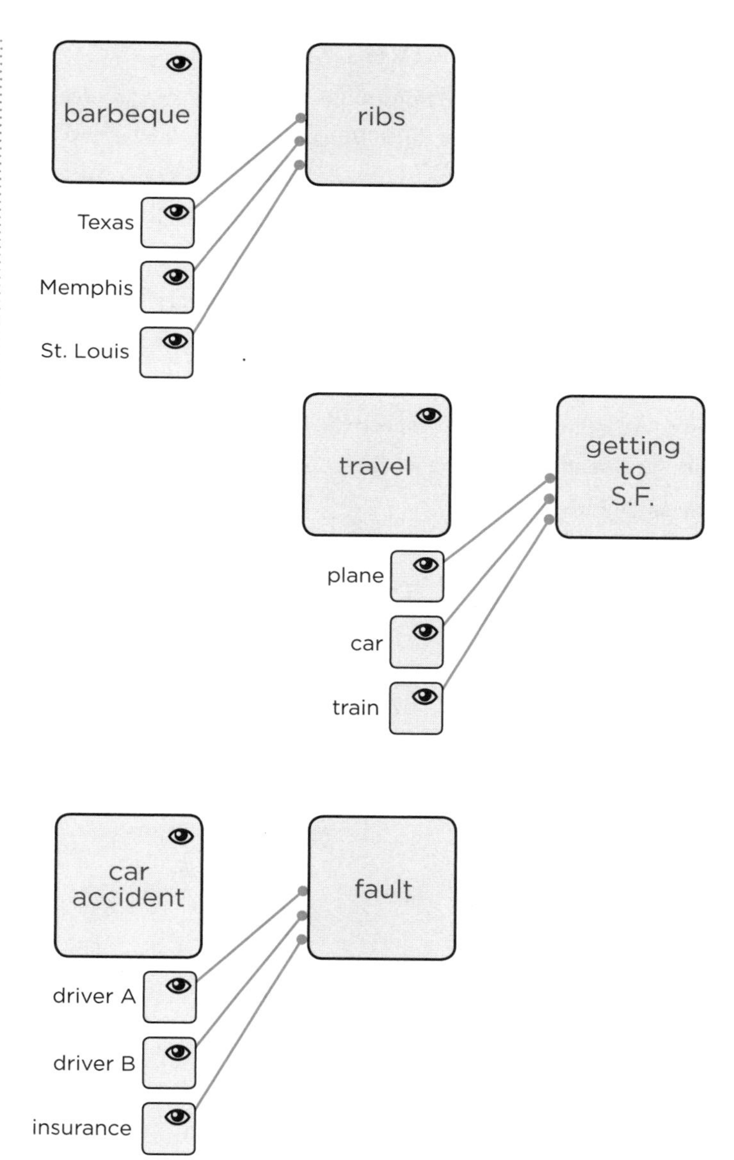

ROOT PERSPECTIVES

Perspectives aren't only from things that have eyes. An idea can be a perspective. For those things for which it is at first difficult to think of as the point of a perspective, think of them as a lens through which you are looking at another thing. For example, we can take a concept as a perspective like "war from the perspective of evolution." Every thought, claim, argument, fact, and idea has a perspective. But often the perspective is hidden: sometimes on purpose, sometimes not. We call these root perspectives. For example, most maps are representations of the world from the perspective of land mass (Figure 5.18).

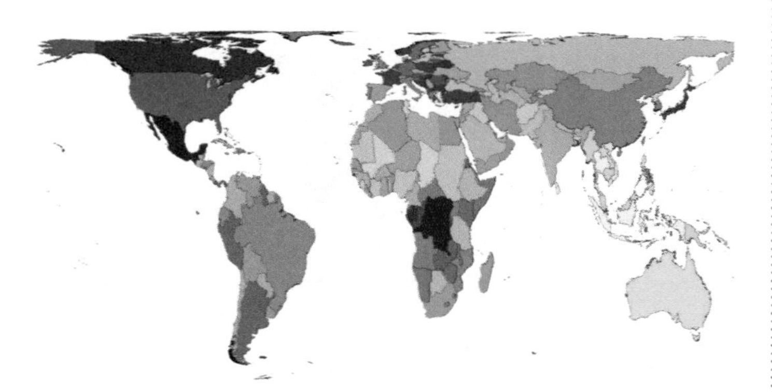

Figure 5.18: Root Perspective of Land Mass

But take a different perspective and you get a very different-looking map. Figure 5.19 shows the world from the perspective of total population.

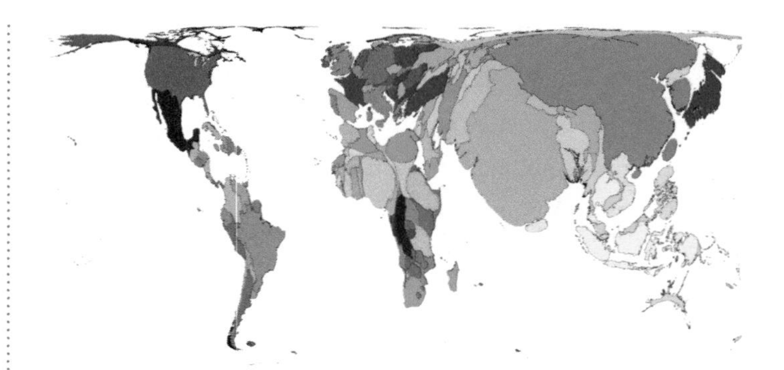

Figure 5.19: Root Perspective of Total Population

More distressing, take a look (Figure 5.20) from the perspective of the incidence of yellow fever[2]:

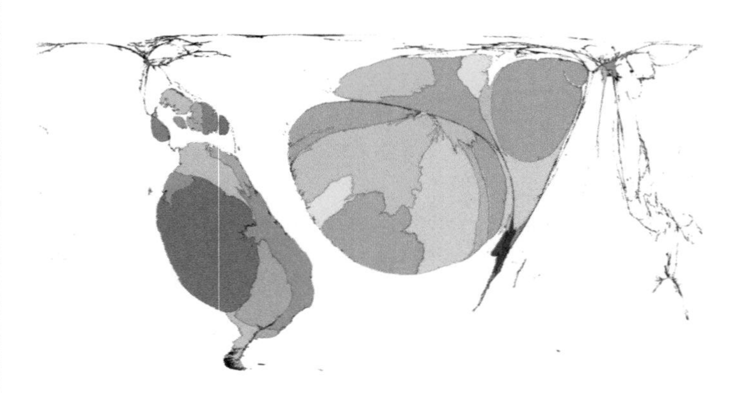

Figure 5.20: Root Perspective of Yellow Fever

[2] All of these maps and more can be found at a fascinating website called World Mapper by network theorist Mark Newman at the University of Michigan. Go to: http://www.worldmapper.org/ © Copyright Sasi Group (University of Sheffield) and Mark Newman (University of Michigan)

As we saw with category *cul de sacs*, there is always a root perspective, even if it has not been made explicit. Developing a mindset that always looks for and finds root perspectives is foundational to critical and independent analysis.

PART-PARTIES

Part-Parties are another cognitive jig that consist of "System and Relationship" get-togethers. We all know that a party where all the guests (parts of the party) stand around and don't mingle (form relationships) is boring. So as we think about our wicked problems it is important to remember to relate all the parts.

Before we discuss Part-Parties, let's clarify something about layout. Note that the maps in Figure 5.21 are all structurally the same. They all show a whole made up of three parts. The way the parts are positioned is a function of layout. For example, if you have parts that have longer words, you might use A or B, whereas C is good for short words, and D is good to show a linear flow read from left to right. The important thing is that you can read these maps, "a whole with three parts."

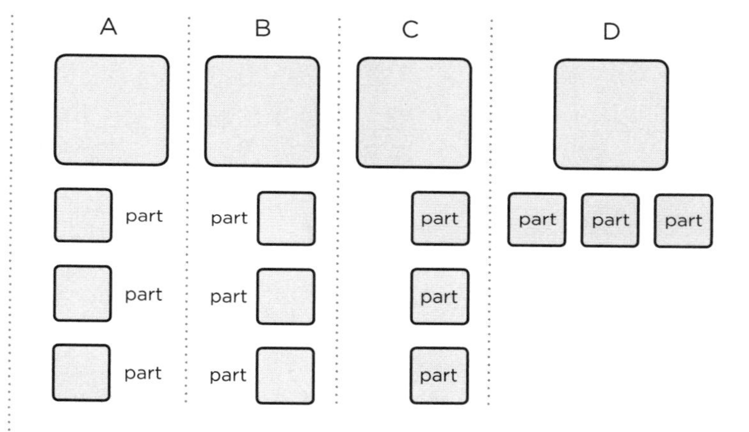

Figure 5.21: Part-Whole Layout

Now back to Part-Parties that are attended by both system (parts of the whole) and relationships (R). It starts with any thing or idea. Let's say that Figure 5.22 is the idea you are going to turn into a Part-Party.

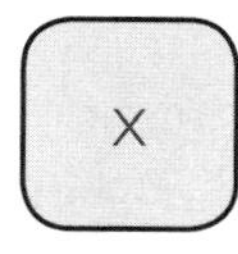

Figure 5.22: Starting a Part-Party

So the basic idea of a Part-Party is (1) break down a thing or idea into parts and (2) relate the parts. A Part-Party is a simple jig that often is overlooked. Figure 5.23 shows four different Part-Parties with 2, 3, and 4 parts, respectively.

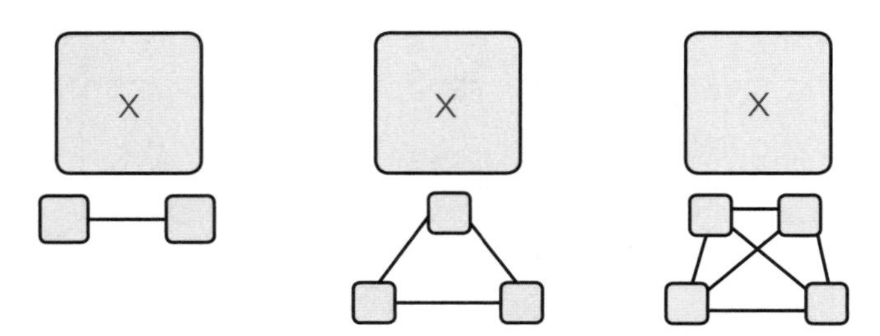

Figure 5.23: Part-Parties with Different Numbers of Parts

Figure 5.24 shows that you can extend the Part-Party by distinguishing the relationships between the parts.

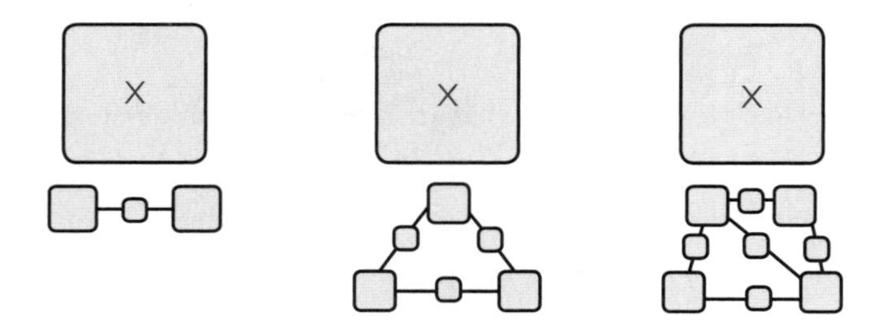

Figure 5.24: Distinguishing Relationships at a Part-Party

Part-Parties can be further extended by including perspective. Much like a real party where everyone has their own unique perspective, all of the ideas or things at your Part-Party also have their own perspectives, as shown in Figure 5.25.

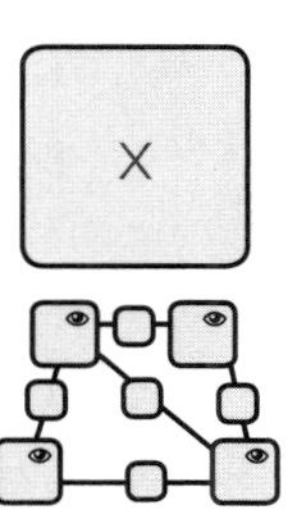

Figure 5.25: Adding Perspectives to a Part-Party

Figure 5.26 illustrates an example of a Part-Party with distinguished relationships and perspectives for three different ecologies. The first (**orange**) is an actual ecological system in Yellowstone National Park, the second (**gray**) is a political ecology at the United Nations, and the third (**tan**) is a word ecology of similar terms.

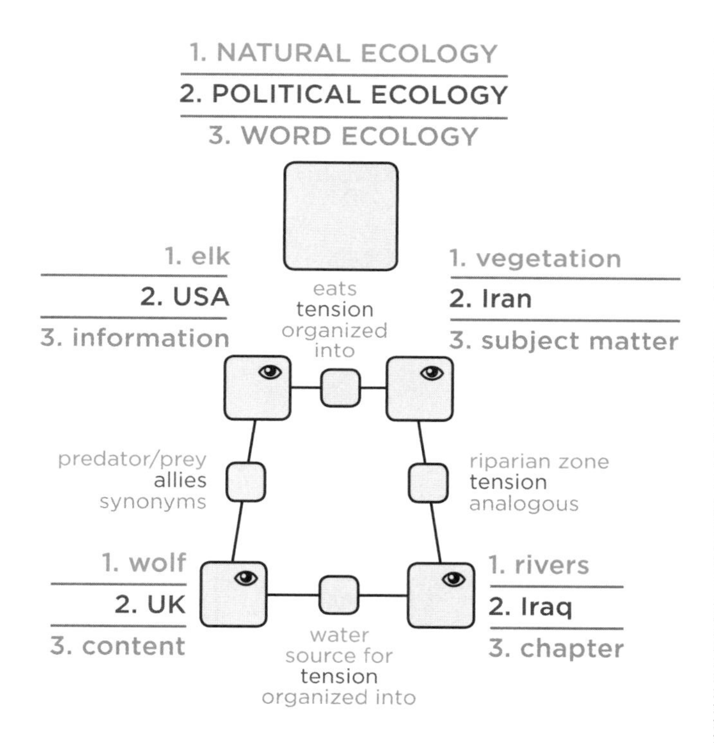

Figure 5.26: Part-Party with Distinguished Relationships

When you learn more about Barbells in the next section, you'll see that the relationships between parts can also become part-whole systems. Combining Part-Parties with Barbells can lead to mapping and modeling of infinite complexity.

BARBELLS

A "Barbell" gets its name for obvious reasons. The basic structure—two ideas or things and the relationship between them—looks like a weightlifting Barbell. All parts of maps that show a relationship between two ideas are referred to as a Barbell, but there are simple and more complicated Barbells. Figure 5.27 shows the simplest Barbell is content agnostic so it can be used and reused for any information or topic. Remember from Chapter 3 (see Figure 3.15) that we use arrows to signify directionality.

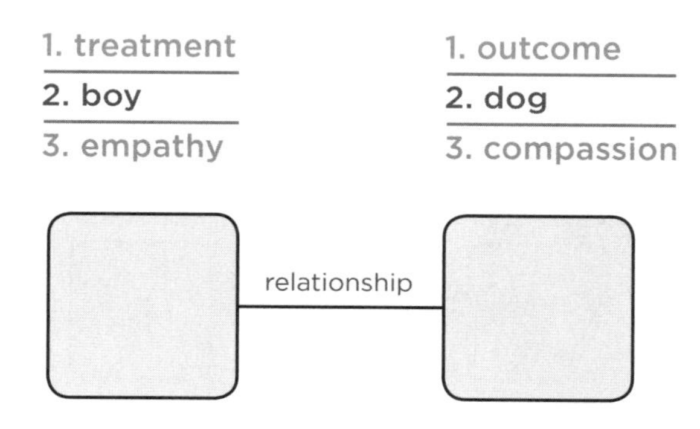

Figure 5.27: Simple Barbell

Simple Barbells can be evolved to become RDS Barbells. These jigs involves three steps shown in Figure 5.28:

- R is for Relate: make a relationship between two ideas or things (i.e., draw a line);
- D is for Distinguish: identify what the relationship is (i.e., add a square and label to the line); and
- S is for Systematize: Recognize the distinct relationship as a system by identifying its parts (i.e., add smaller squares).

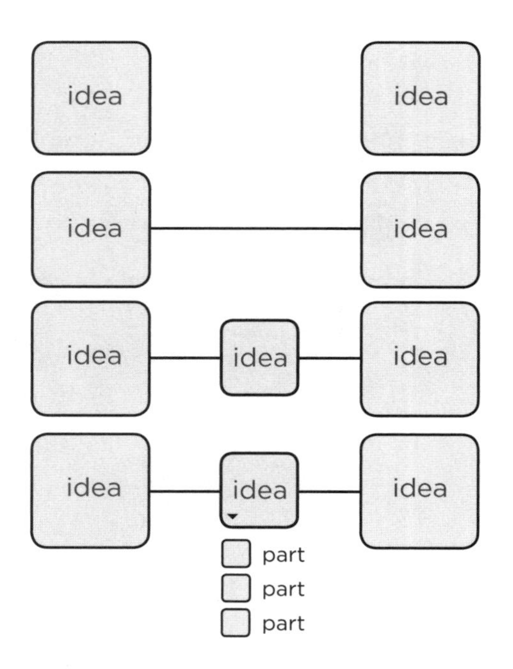

Figure 5.28: The Steps of an RDS Barbell

RDS Barbells are extremely important jigs because much of the mystery and complexity of wicked problems is hidden in the interrelationships between ideas or things. RDS Barbells guide us toward identifying and deconstructing this complexity.

Figure 5.29 shows that the same RDS Barbell can be used for feedback loops.

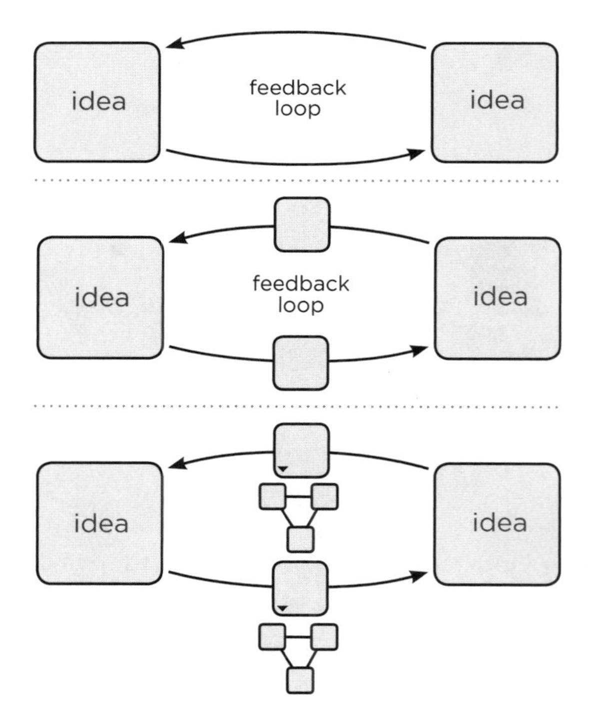

Figure 5.29: RDS Barbell Showing Feedback

We sometimes call RDS Barbells algorithms for innovation because they can be powerful ways to generate new ideas. There's a saying that there are really only three ways to innovate: (1) invent something totally new, (2) make an existing product better, or (3) combine two existing things into something new. For example, RDS Barbells illustrate the innovation process needed to think of quickmarts (the relationship between a gas station and a food market) and biochemistry (the interdisciplinary relationship between biology and chemistry).

RDS Barbells are useful any time you want to relate (d) parts (b and c), allowing you to explore that relationship even further. Figure 5.30 illustrates that an RDS Barbell can occur anywhere, even as parts to a whole.

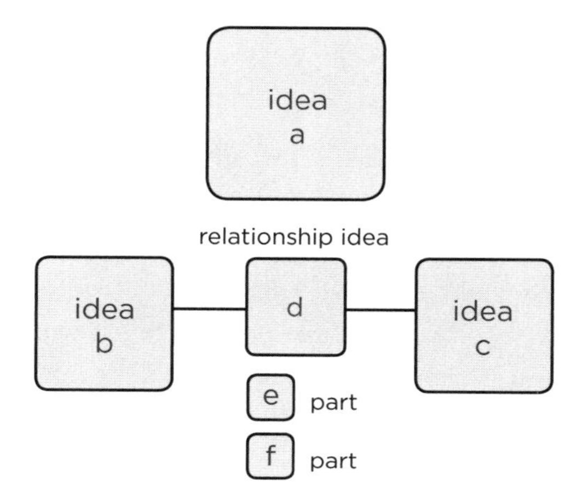

Figure 5.30: RDS Barbell as Parts to a Whole

Or this jig from Figure 5.30 could represent the concept of interaction design shown in Figure 5.31, where interaction design is made up of the parts user (b) and experience (c), which are related through an interface (d) that is made up of two parts, hardware (e) and software (f).

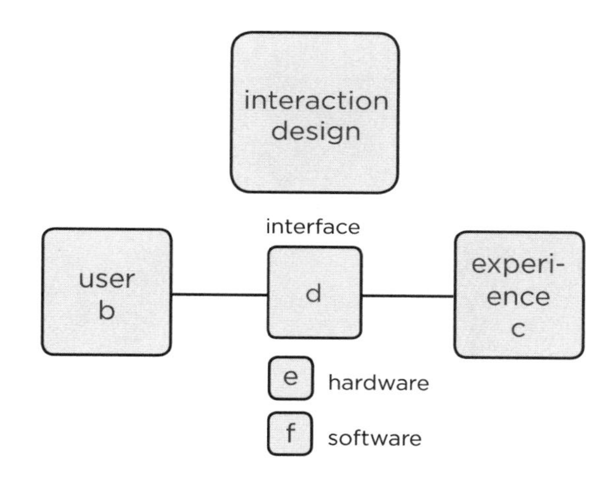

Figure 5.31: Example of RDS Barbell

Or consider the same jig structure in Figure 5.32, which illustrates a doctoral dissertation (a) consisting of the relationship between non-marital births (b) and the 1996 Welfare Reform Act (c). This relationship was found to be correlational (d) based on two parts, mediating variables (e) and legislative provisions (f).

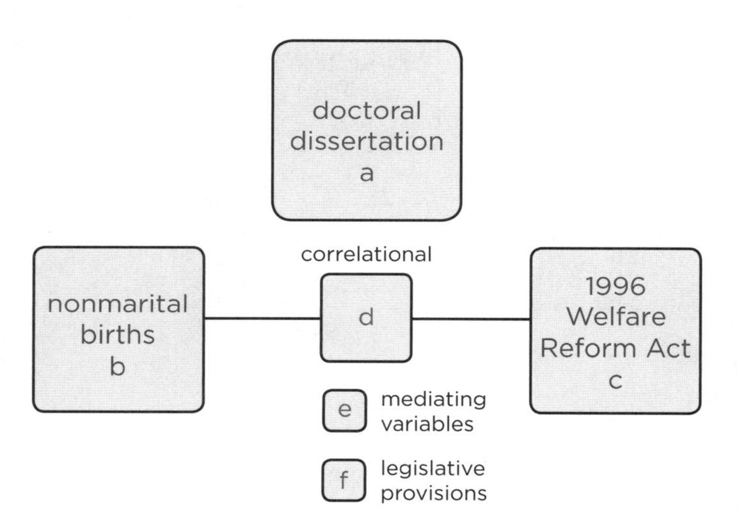

Figure 5.32: RDS Barbell Example 2

STRINGING BARBELLS TOGETHER

Mapping of more complex systems is accomplished simply by stringing jigs such as directional Barbells, RDSs, and R-Channels together. We can also position things in different ways to create relationships between more than two ideas, as in Figure 5.33.

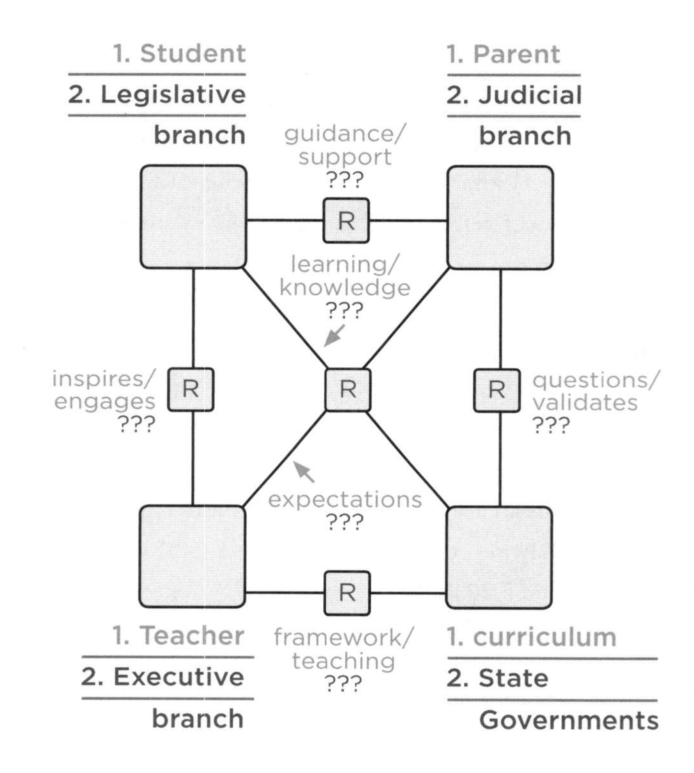

Figure 5.33: Increasing Complexity with Barbells

The example in Figure 5.33 illustrates Barbells strung together. The content examples include deconstructing the interrelationships between the 3 branches of government (**gray**) or between student, teacher, and curriculum (**orange**). Note that a "?" shows that maps can be used equally to organize known information, and to generate new information as part of the teaching and learning processes. Take a minute to think about how you would label the relationships marked with a "?".

The level of complexity that can be modeled is effectively infinite. Figure 5.34 illustrates how RDS Barbells can be nested inside of RDS Barbells. The point is that DSRP simple rules and jigs can expand your systems thinking. Be creative!

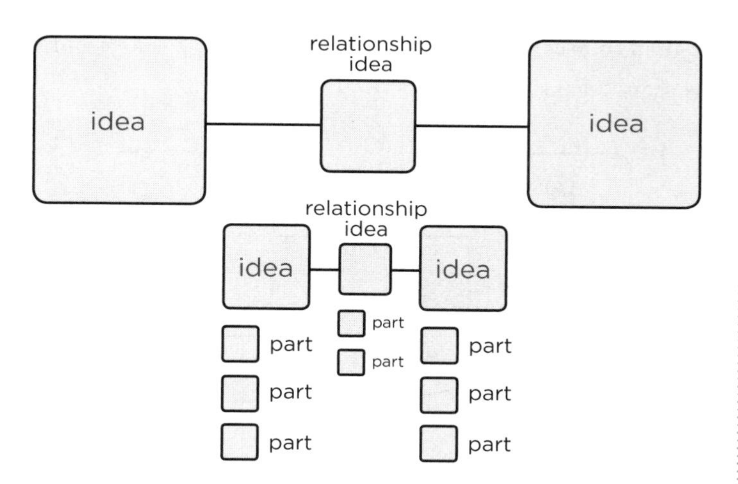

Figure 5.34: Complex RDS Barbell Inside of an RDS Barbell

R-CHANNELS

Next we will discuss a subtype of RDS Barbells called R-Channels. Figure 5.35 shows that often the ideas being related also have parts.

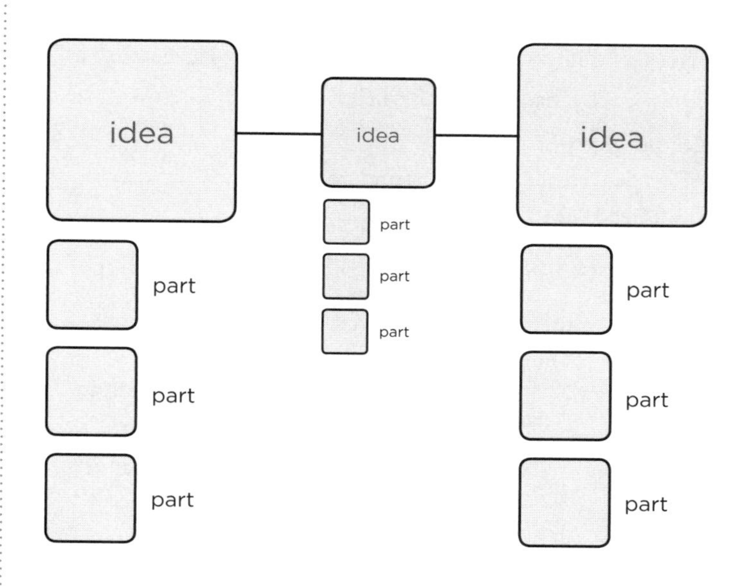

Figure 5.35: RDS Barbell Between 2 Systems

An R-Channel (Relationship-Channel) is a jig that opens up a channel between two systems made up of parts so that we can relate not merely two ideas or things, but all of their parts. So it uses the jig in Figure 5.35 and left and right justifies the parts so that a channel is formed for relationships between them. For example, we might relate the parts of a specific medical treatment with the parts of its outcomes. Figure 5.36 shows an R-Channel. The red lines show the "channel" that is caused by justifying the parts left and right, thereby making a space between them in which to draw relationships.

Figure 5.36 illustrates the power of combining jigs because it shows an R-Channel with Barbell Relationships between parts.

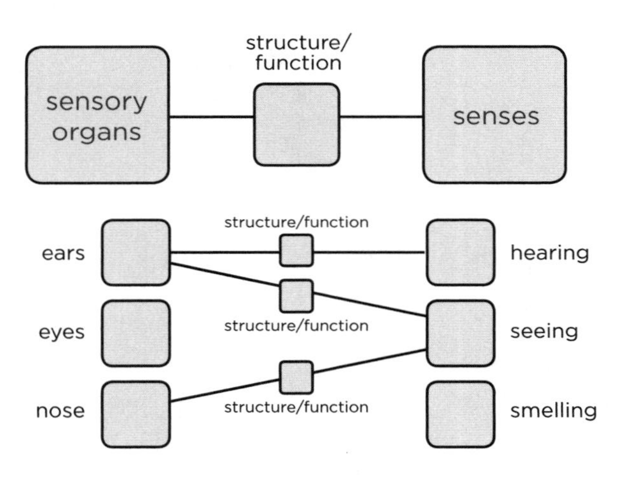

Figure 5.36: Part-to-Part RD Barbells: Example 1

Figure 5.36 shows us that the relationships between all the parts (x1, x2, and x3) are all in turn part of the larger relationship (x). On flat paper we have to imagine this, but in MetaMap software we can do this by adding distinctions to each relational line and then dragging those onto the relationship in the original Barbell. The content examples provided include a prior learning assessment (PLA) student matching their life experience and competencies to the required courses in order to get a degree and the 5 senses and sensory organs.

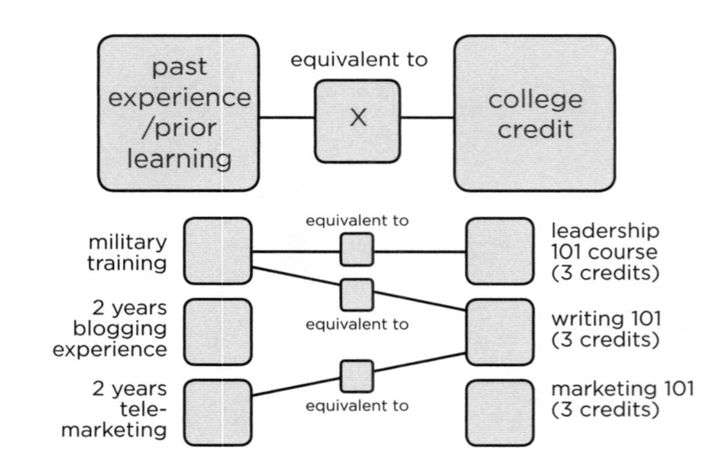

Figure 5.37: Part-to-Part RD Barbells: Example 2

CHAPTER 6

MAKE STRUCTURAL PREDICTIONS

PREDICTIVE POWER OF SIMPLE RULES

To predict the future is the ultimate goal of all science: to understand the intricacies of the universe well enough to make predictions of what will occur. Prediction: if you smoke, you will greatly increase your odds of getting cancer. Prediction: the force will equal the mass times its acceleration.

DSRP rules also help us to make predictions about the structure our mental models (knowledge) will take, predictions about how we will think. For example, we can predict that if two ideas are located in the same conceptual space we'll try to relate them.

Figure 6.1: Relationship?

This prediction lies at the base of marketing, where we are constantly bombarded with free associations that benefit the manufacturer. We can effectively predict that you'll think yellow and fruit.

For example, we can predict to a profitable degree that children relate their favorite sports star (Ken Griffey, Jr., Reggie Jackson, or Nolan Ryan) with Frosted Flakes cereal (a spurious relationship). And large marketing budgets have been expended to associate fat free with healthy (also a spurious relationship), so by triangulation: fat free = healthy, frosted flakes = fat free, therefore frosted flakes = healthy. That's all that kids and moms need to make the purchase and begin the morning with a transfusion of 21 grams of sugar per bowl.

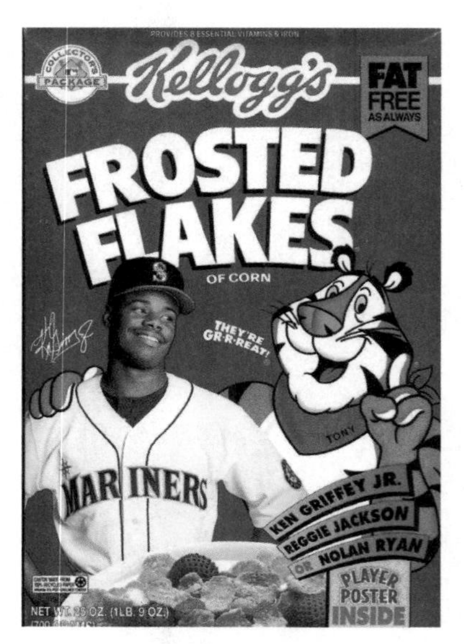

Figure 6.2: Frosted Flakes Is Healthy

Coca Cola has engaged us all in a decades-long campaign to create an ecology of interrelated ideas that lead us to believe that a Coke should be enjoyed during all of life's moments. Despite the fact that Coke actually leads to an early death,[1] Coke is spending hundreds of millions to ensure that children grow up believing that Coke and life are intertwined. They've also ensured that you'll draw a relationship between Coca Cola products and: Santa Claus, happiness, the United States of America, and friendship. That's the power of prediction based on your mind's natural processes.

Figure 6.3: Coke Is Friendship and Happiness

Marketers and political operatives know that if they cause the general public to make any given relationship enough times many will continue to make it even when it's not being made for them. Eventually, the relationship between two distinct things or ideas soon merges into a single system—a new thing/idea comprised of the interrelated parts. The word "is" is the third person singular present form of the verb to be. To be is an existential idea—it gets at the very essence of something. These manipulations are intended to mess with our distinctions. Frosted Flakes is healthy. Coca Cola is life. Obama is Osama.

Figure 6.4: Obama Is Osama

This rewiring of our most basic distinction-making is predictable. But prediction need not only be used for profit and manipulation. It can be a powerful inoculation against such manipulations. If we can predict the structure our own thinking will take, it can be used to discover new ideas.

PREDICTION IN KNOWLEDGE CREATION

One of our biggest systemic problems is that we are unable to keep up with the sheer number and magnitude of wicked problems we face. It seems life is speeding up with globalization. This means we need to create knowledge faster just to keep up with all our wicked problems. Nowhere is this more evident than in the fight to keep up with evolving superbugs

[1]Aune, D. (2012). Soft drinks, aspartame, and the risk of cancer and cardiovascular disease. *American Journal of Clinical Nutrition*, 96(6), 1249-1251.

(strains of bacteria that evolve to become resistant to our antibiotics), where the evolution of these bacteria is out-pacing the evolution of our knowledge about them. That means we need to not only create more knowledge (new discoveries), but create that knowledge faster. Up until recently, our knowledge creation industry (mostly scientific labs in universities and innovation labs in organizations) has kept up with the demand reasonably well. But as demand increases, supply is hampered by three factors: (1) a lack of understanding of how knowledge is created, (2) over-focus on informational content, and (3) inability to be interdisciplinary.

In Chapter 4 we discussed that meaning comes from the interaction of information content and DSRP structure. Let's look at a very simple example in Figure 6.5 using just a few words. We can think of the words themselves as information content and any shapes or lines as being the structure, so in the image below, the words Jump, Dog, and The are bits of information. The only structure they have thus far is that they are distinctly different because they are different squares. Notice that, thus far, there is very little meaning being created because we haven't structured this information. Sure, we know what the words themselves mean from prior knowledge, but we have no idea what the meaning of the words are here together in this context.

Figure 6.5: Unstructured Information

Now, let's add some structure, as seen in Figure 6.6. You can see that by simply putting the three distinct ideas in a row and relating them, we get, "Jump The Dog." We now have some meaning being made. Apparently there is a dog and someone is giving a command to jump over it.

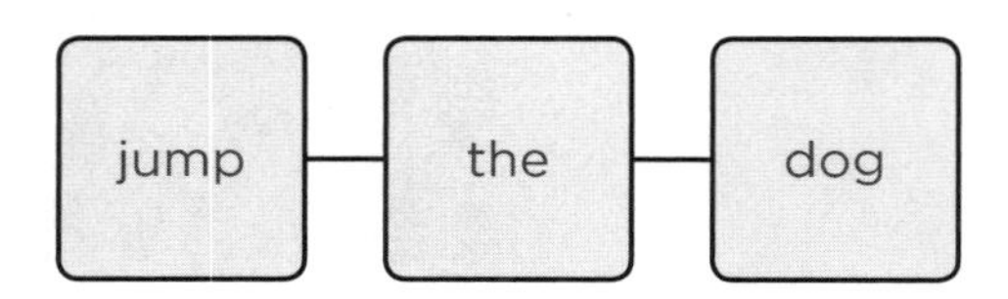

Figure 6.6: Ordered and Related Information

But if we change the organization or structure, we get a very different meaning. In Figure 6.7 we see the configuration: "The Dog Jump." Forced to make meaning out of this and only this structure, we might come to the conclusion that there's a cool jumping ramp for dogs, or something like that.

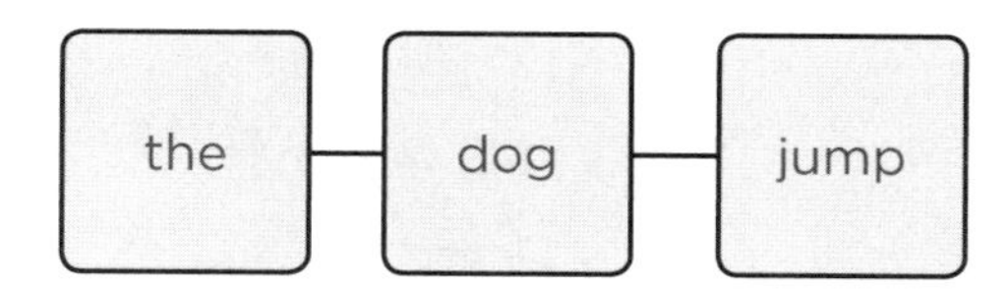

Figure 6.7: Re- Ordered Information

The point of this simple example is to help you differentiate between information and structure. The structure is all the squares and lines, and the information is the words themselves. Structuring information is what thinking is. Thinking is a process of structuring information in order to give it meaning. So if you take the same information but structure it differently, you'll get different meaning. Likewise, if you take the same structure, but put in different information, you'll get different meaning.

What DSRP does, using a form of metacognitive mapping which we can best exemplify through MetaMap software, is to show you all the possible ways the information you are working with *could* be structured. You may choose not to structure it in some of the ways that the DSRP and the software suggest, but you could. When you do follow the predictive power of metacognitive mapping using DSRP, you'll find that you are able to create new meaning and garner new insights about whatever it is you're thinking.

In addition, because DSRP (and MetaMap software) predict using patterns of thinking that are universal, it naturally guides you to thinking that is more complex, more robust, more complete, and more systemic. That is why DSRP is referred to as systems thinking.

Interdisciplinarity is important because wicked problems do not respect disciplinary boundaries. So our old disciplinary approach—where a physicist solves physics problems and a biologist solves biological problems and an economist solves economic problems—won't do. Problems don't say to themselves, "I'm just an economic problem, I'm not going to mess around in engineering or chemistry or biology." Our lack of understanding of how knowledge is created is deeply rooted in our excessive focus on and allegiance to informational content over cognitive structure. All three of these problems of knowledge creation can benefit from application of DSRP simple rules.

Let me give you two examples that come from different types of knowledge creation. The first is the knowledge creation of an individual group of first graders. Of course, the knowledge they are creating isn't new knowledge in the sense that it's not new to society as a whole, but it's new to those kids. The second example is new knowledge, globally speaking: the discovery of something completely new and groundbreaking. Both examples will illustrate why recognizing the often hidden structure of DSRP rules is so important to the process of creating new knowledge.

A first-grade teacher we worked with committed to teaching her students just one of the simple rules: part-whole. She did a short five-minute lesson where she explained that "everything can have parts and the parts can have parts." She played a few fun games and led the kids in song about "parts and parts and parts."

She then had the kids do what she had done every year for five years: go out into the parking lot and meet some firemen and see their firetruck. After the field trip, the kids drew and labeled pictures of their firetrucks and built a firetruck (they can sit in) out of cardboard boxes, paint, and other stuff. What this teacher found was that with just a five minute lesson on one of the universal structures (Systems Rule or part-whole) the kids drew more detailed firetrucks, they used more vocabulary words to describe their firetrucks, and they built a more sophisticated cardboard firetruck with more parts and distinctions! An example is shown in Figure 6.8.

The teacher explained that in previous years, where a pre-lesson on part-whole was not included, children's drawings included far fewer parts, usually the basics like front, back, steering wheel, ladder, and tires (see Figure 6.9).

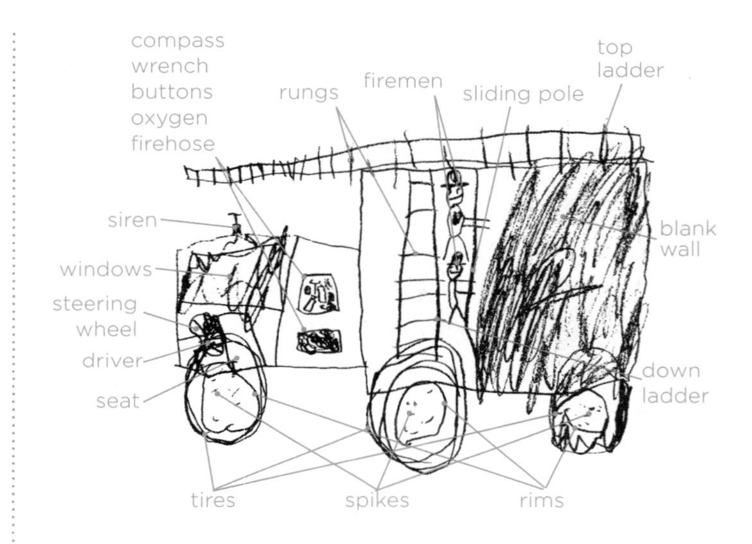

Figure 6.8: Post-Parts Lesson Firetruck

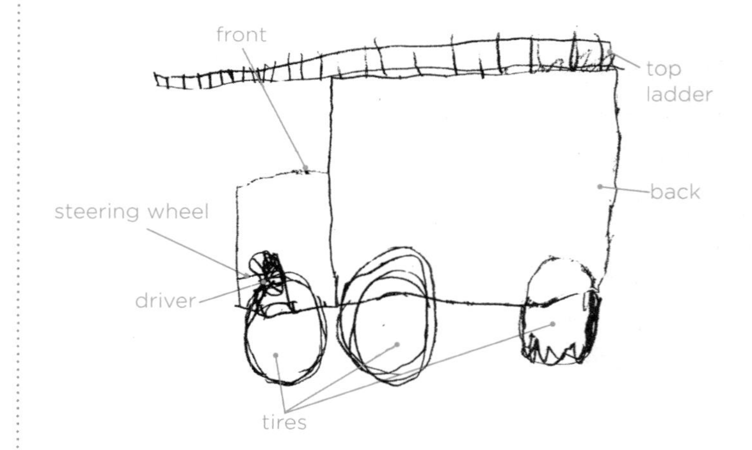

Figure 6.9: Pre-Parts Lesson Firetruck

Later that semester, the same kids sent us a note (we had visited their classroom several times during the study) describing how they went on a field trip to an apple orchard. While there, they started thinking about all the parts of an apple and discovered one that they didn't have a word for. Maybe no one had a word for the little fuzzy part, they thought. Maybe no one has even *discovered* ever before the little fuzzy part! They wanted to know if we, as scientists, knew if it had ever been discovered. And, in case it hadn't been, they had a suggested name: the apple's *bellybutton*. We let them know that it had already been discovered. After their initial disappointment, they proudly began using the term *calyx* in sentences.

A month later in the semester, the school went into lock-down mode because of something that happened in the wider community. For first graders, experiencing a lock-down for the first time can be a traumatic and worrisome experience. After it was over, a little girl asked her teacher, "Ms. Ward, can we part-whole the lock-down?" The whole class made a part-whole diagram of the lock-down to better understand what had just happened.

What's remarkable is that these kids, without anyone asking them to, accomplished what educational scientists call *far transfer*. Far transfer means that when you learn something in one domain, you can transfer it to another domain and use it there. The term far refers to the degree of relatedness of the domains. Far transfer means that when someone teaches a student one thing, they can teach themselves 5, 10, or 20 things. What's truly remarkable is that firetrucks, apples, and lock-downs have very little in common from an information-content perspective, but structurally, they were similar for these kids, and they knew it. The point is this: what transferred for these kids was not the content of the lessons (firetrucks, apples, lock-downs) but the structure (part-whole).

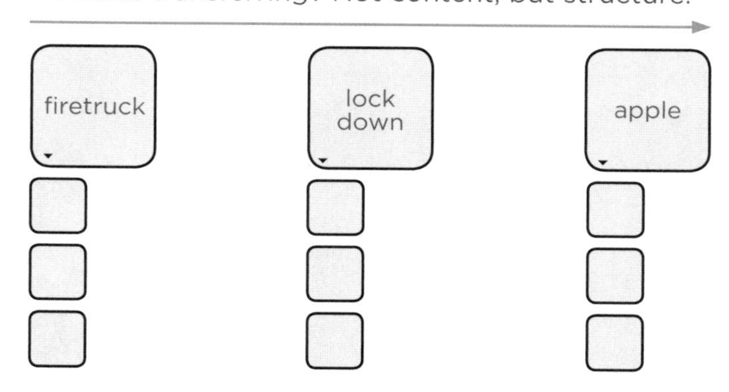

Figure 6.10: Far Transfer

That's all well and good for gradeschoolers, but what about knowledge creation for society as a whole? Charles Darwin took two books with him on his famous voyage on the *Beagle*. One was the Bible and the other was Sir Charles Lyell's *Principles of Geology*. The latter was instrumental for Darwin in discovering his profoundly important ideas about evolution.

Darwin saw a *structural* analogy between geological and biological uniformitarianism. The content of what Darwin read in Lyell's book (geological stratification) was substantively unlike the things Darwin was considering (biological speciation), but the underlying structure was, to Darwin's mind, very similar. Here again, we see the immense importance of cognitive structure in the formulation of new knowledge. Time and time again, great inventions and innovations are the product of this structural awareness rather than informational content.

What's transferring? Not content, but structure.

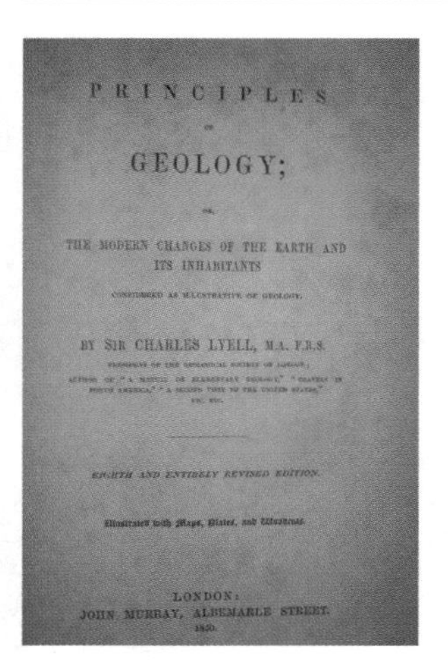
PRINCIPLES

GEOLOGY;

THE MODERN CHANGES OF THE EARTH AND ITS INHABITANTS

BY SIR CHARLES LYELL, M.A. F.R.S.

EIGHTH AND ENTIRELY REVISED EDITION.

LONDON:
JOHN MURRAY, ALBEMARLE STREET.

geological uniformitarianism

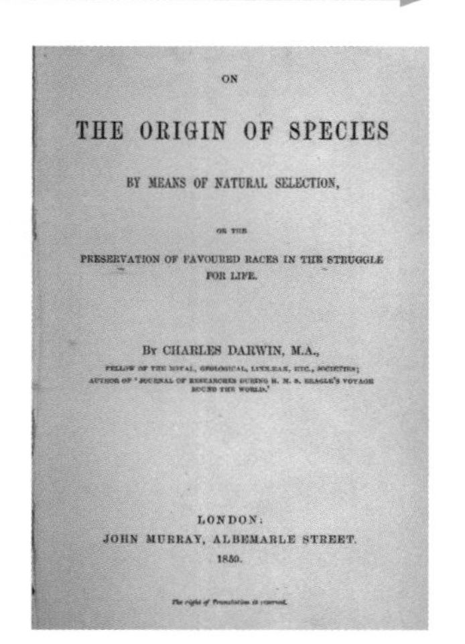
ON

THE ORIGIN OF SPECIES

BY MEANS OF NATURAL SELECTION,

PRESERVATION OF FAVOURED RACES IN THE STRUGGLE FOR LIFE.

BY CHARLES DARWIN, M.A.,

LONDON:
JOHN MURRAY, ALBEMARLE STREET.
1859.

biological uniformitarianism

Figure 6.11: Uniformitarianism in Geology and Biology

Remember from Chapter 4 that many traditional mapping techniques emphasize informational content over structure. In contrast, metamaps emphasize the structure that brings meaning to information. Knowledge discovery in the future may very well be reliant on the big data created by these globally shared metamaps that show us where new knowledge can be discovered based on structure, not on substantive content.

The University of California at San Diego's (UCSD) Map of Science (Figure 6.12) illustrates the analysis of 7.2 million papers (sourced from Web of Science and Scopus) using author citation networks across 13 sub-disciplines.[2] David McConville of the Buckminster Fuller Institute explains: "The radical fragmentation of knowledge is making it difficult to understand any kind of big picture. Academia now has 8,000 disciplines, 50,000 journals, and over a million articles published every year. A visualization from the University of California–San Diego shows how few disciplines actually draw from, or even reference, fields other than their own. We are facing a time of extreme hyperspecialization."[3]

[2]Börner, K., Klavans, R., Patek, M., Zoss, A. M., Biberstine, J. R., Light, R. P., Larivière, V., Boyack, K. W. (2012). Design and Update of a Classification System: The UCSD Map of Science. *PLoS ONE* 7(7): e39464. doi:10.1371/journal.pone.0039464

[3]Fideler, D. (2015, January 7). Putting the World Back Together: The Future of Education and the Search for an Integrated Worldview. Retrieved March 23, 2015, from http://www.cosmopolisproject.org/2015/01/07/putting-the-world-back-together-the-future-of-education-and-the-search-for-an-integrated-worldview/

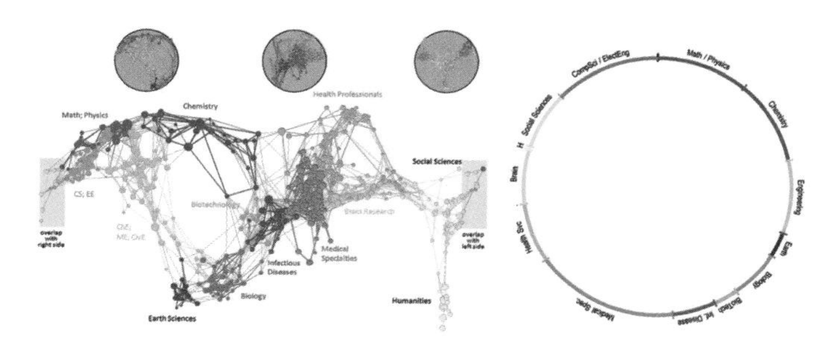

Figure 6.12: Science Is Still Disciplinary

This lack of interdisciplinarity equates to a lack of innovation in science. The kinds of profound insights had by Darwin and children alike come when disciplines intermingle and underlying structures are explicated. Systems thinking and interdiscplinarity are synonymous sister acts in moving toward a new science. This new science is based on DSRP structures because it is: boundary crossing (D), reductionistic and holistic (S), interconnected and relational (R), and hyper-perspectival (P).

Simple DSRP rules lie at the base of much of what we need in order to build interdisciplinary knowledge faster and better and ward off the wicked problems of tomorrow, because they:

1. Give us a modeling language for understanding how knowledge is created;
2. Reduce excessive focus on informational content by balancing it with cognitive structure;
3. Increase interdisciplinarity and transfer;
4. Give elite scientists with highly specialized terminology a common language; and
5. Provide predictive analytics that allow scientists to "mine for insights" in the most probable places.

USING IDEA ANALYTICS TO PREDICT NEW IDEAS

The predictive capabilities of DSRP are as simple as the theory itself.

You can learn to use these predictive powers in just minutes.

DSRP is a basis for metacognitive mapping that is powerfully predictive, it can help people in any field or situation think better about anything.

The way DSRP does this, paradoxically, is not to focus on the topic that one is thinking about. The topical details are irrelevant to the predictive power of DSRP. DSRP helps one focus on predicting the structural possibilities of ideas, rather than the content information that maps onto those structures.

The predictions that DSRP makes are analytics. There are two distinctions that are important to make to give us a framework for the analytics that DSRP provides. First is a distinction between a whole metamap (a map) and a single idea (a thing). We'll call these respectively map analytics, which gives you the analytics for the entire map, and thing analytics, which gives you analytics for an individual idea/thing. The next distinction to make is between what you have done and what you could do. We'll call this have done and could do and plot these in Table 6.1. Have done means that you have already followed the particular DSRP rule for the thing or map, whereas could do refers to all the possible structures that you could create if you wanted to. Have done is descriptive, whereas could do is prescriptive because it will help you to generate new ideas.

Table 6.1: Important Distinctions in Analytics

	Thing	Map
Have Done (descriptive)	The DSRP structures you have already taken into account for a given thing or idea	The DSRP structures you have already taken into account for the entire map
Could Do (prescriptive)	The DSRP structures you could take into account for a given thing or idea	The DSRP structures you could take into account for the entire map

Let's look at how they work using metmap diagrams and coloring because it's easier to visualize and explain, but it is important to note that you can also just use DSRP conceptually, in your mind, without the the metamaps. Remember that DSRP is associated with four colors shown in Figure 6.13, which illustrates what we have learned thus far in the book, which is that any idea or thing can follow any one or all of the DSRP rules at any time, thereby producing new structures.

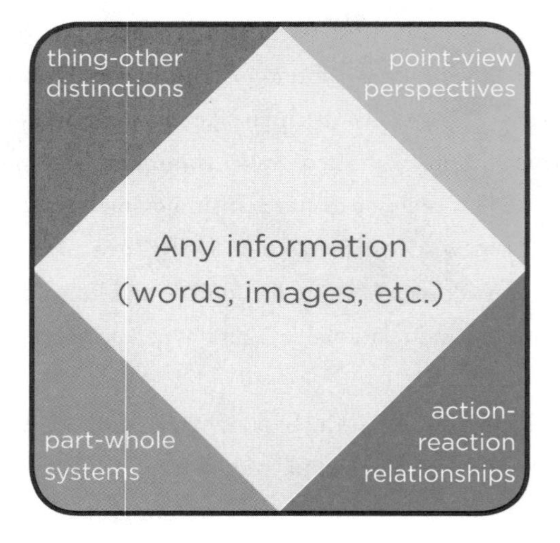

Figure 6.13: Any Thing or Idea Can Follow DSRP Rules to Create New Structures

These colors will help us quickly see where we have executed the DSRP rules and where we haven't for each idea or thing and for the map as a whole. Figure 6.14 illustrates a simple map with some of the descriptive "have dones" identified by color for not only each idea on the map, but for the map as a whole. We use an example of a political ecology but because we are not analyzing content, it is the structural properties of the map that are most relevant.

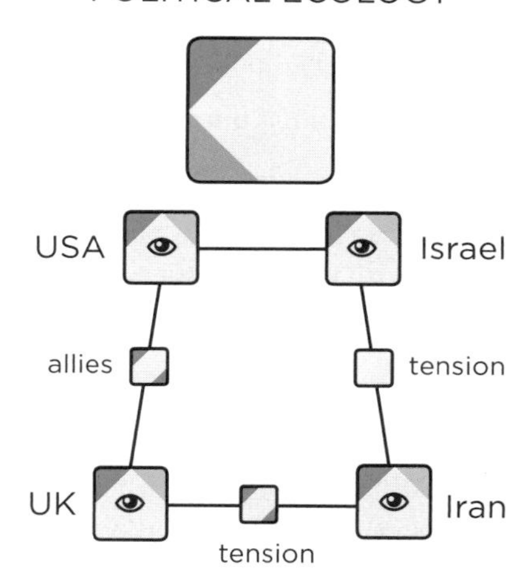

Figure 6.14: Descriptive Have-Dones for Each Thing or Idea and for the Map

Note that there are 8 distinct identities or things on the map. Only 1 of those have been converted into systems by adding parts, 4 relationships have been made and 4 perspectives have been taken. Right away we can begin to see room for improvement in our thinking. But zooming into any individual thing or idea, we can see which rules have been used and which ones have not. For example, red indicates that the distinction rule has been used, so anything that has been created will receive a red indicator. A green indicator means that that thing has been converted into a system by breaking it down into parts. If no green indicator exists, then the opposite is true, it hasn't been broken into parts but it could be. The same applies for the blue indicator for relationships and the orange indicator for perspectives. You'll notice that 4 relationship have been made in the map analytics, but 7 things in the map are affected by those 4 relationships and therefore express the rule: 4 ideas are related and 3 ideas are distinguished/identified relationships. One relationship exists which has not been distinguished but could be. The 4 countries are acting as perspectives on each other.

Figure 6.15 shows the opposite map: prescriptive or could-do analytics. Notice that the map analytics have not changed: 8 distinct ideas, 1 part-whole system, 4 relationships, 4 perspectives. The colors, however, have flip-flopped. These analytics indicate what could be done prescriptively or generatively to create new structures and have new ideas. A green indicator means that the thing or idea could be broken into parts. A blue indicator means the thing or idea could be related to other things or ideas. A red indicator (see top relationship line) means the relationship could be distinguished (and subsequently systematized, etc). An orange indicator means the idea or thing could be the point or a view of a perspective.

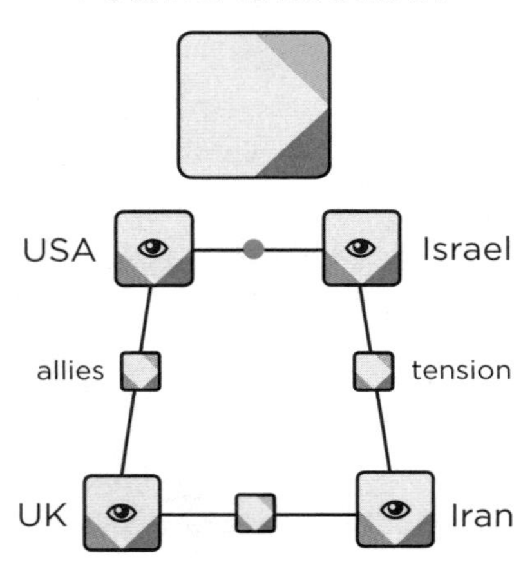

Figure 6.15: Prescriptive Could-Dos for Each Thing or Idea and for the Map

There are numerous other analytics that can be viewed individually per idea or collectively for the map. For example, the thing analytics for the top square labelled political ecology in Figure 6.15 is shown below in Figure 6.16. Note it is distinctly different from 7 other ideas (there is also an advanced distinction analytic which has to do with explicitly distinguishing ideas). It is not a point but is the view from 4 perspectives. It is part of nothing but contains 7 parts (3 of them relational). It is neither related to nor acting as a relationship. The thing analytics image to the left provides a summary: the thing labelled political ecology is 1 of 8 distinctions (⅛), part of nothing but having 7 parts of its own (7/0), it is neither a point nor a view (0/0), and is neither related to or, relating (0/0). In all the places where there is a 0 in the analytics, there is room for thinking that could be done. Of course, even where there are natural numbers (i.e., 1,2,3…) there is room for improvement. We could for example consider which system political ecology is part of such as human ecology that might include social, economic, and technological dimensions. Likewise, there are 196 countries in the world that could all be parts alongside the four we mentioned. Each of these countries is not a homogeneous whole but would have political factions within, so they could be further broken down into these parts.

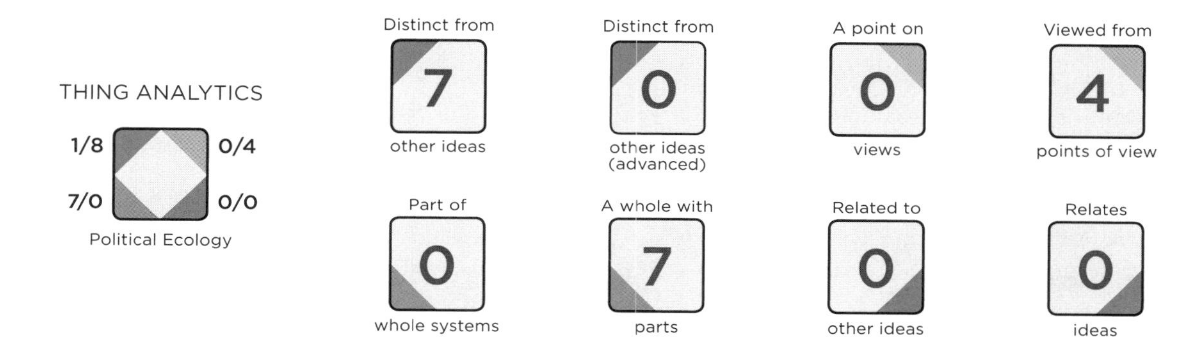

Figure 6.16: Analytics for a Single Thing or Idea

In Figure 6.17 you can see the map analytics for the map in Figure 6.16 showing analytics that combine structures. For example, there are no viewpoints and therefore no viewpoints have been made into systems, but there could be. There are 4 relationships, only 3 relationships were made into distinct ideas, and of those none have been made into a system or a perspective. Combining DSRP analytics further provides insight. There are 8 ideas in this map, yet only 1 of them have been broken into parts. Likewise, only 4 relationships of the 6 possible between the parts have been made.

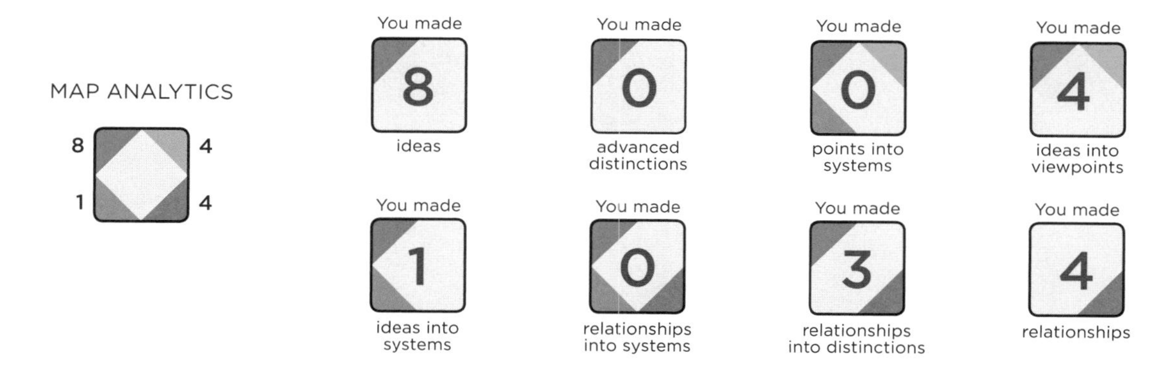

Figure 6.17: DSRP Combinations Analytics for a Map

This might seem like a simple set of predictions, but realize that you can follow the predictive analytics for all ideas and all relationships in your maps. So, for example, if you have 5 or 10 basic ideas making up your thinking, you would be able to select on any one of them and see the predictive analytics. As you followed one or more of those suggested paths, with each addition of an object or relationship you'd have more options. In addition, the predictive analytics are numerical, so even though there is one analytical slot for creating parts, you could create multiple parts, increasing the number in the analytics slot. The same is true across all of the numbers in the analytics. This is, of course, something that you can keep track of in your head simply by following the four simple DSRP rules. But our research lab also programmed these analytics into MetaMap so that one can deal with more complex analyses.

MetaMap is a powerful web-based software that guides you through its predictive analytics. It provides a real-world, practical, and visual example of what we mean when we say that DSRP is predictive. DSRP (with or without the software) predicts the future possibilities of systems thinking by predicting the structures that are possible for any idea about any topic.

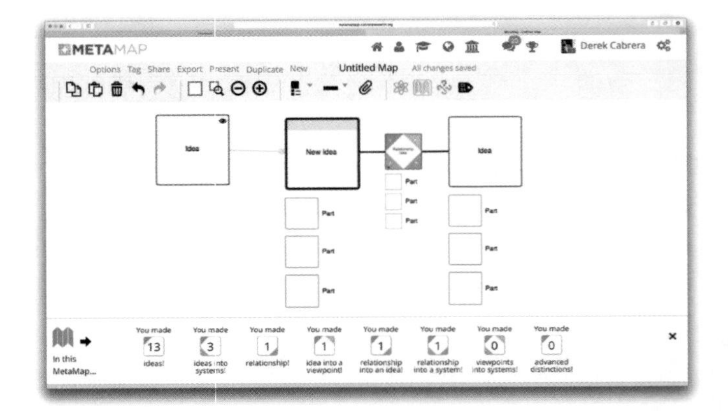

Figure 6.18: Predictive Analytics in MetaMap Software

Let's look at a substantive example of the predictive power of DSRP. Recent research[4] identified the primary components of the food-water-energy nexus, depicting (Figure 6.19) the need to see these three vital resources as undeniably connected when looking for a solution to global resource management as the population grows.

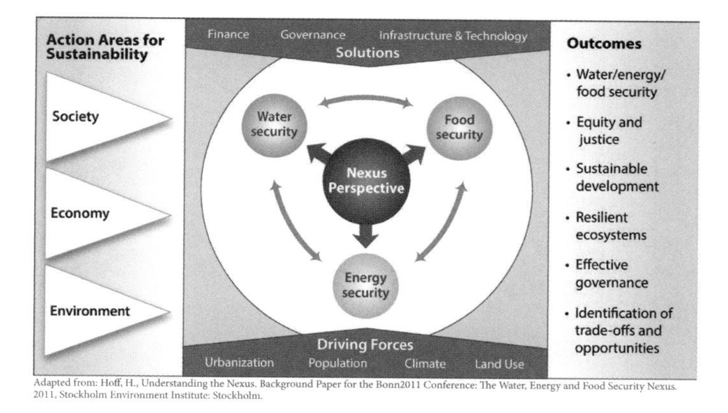

Figure 6.19: The Water, Energy, and Food Security Nexus

Figure 6.19 serves as a good outline of the major ideas that researchers see as inputs, factors, and outputs when explicating what they call the "nexus perspective." Note the delineation of four items—Action Areas for Sustainability, Solutions, Driving Forces, and Outcomes—all surrounding the graphic of the food-water-energy nexus. Note also that the center graphic explicates the three primary components (food, water, and energy) that the nexus perspective must account for when offering insight into an impending resource shortage.

The predictive power of DSRP tells us that we must thoroughly examine each of the components of this idea of a water-food-energy nexus, and provides us with an algorithm to predict where and how we must dig deeper to truly understand and act upon these ideas.

For example, to focus on the structure of the nexus, the metamap in Figure 6.20 is a reproduction of the elements in Figure 6.19. This map shows both the content and the underlying structure of the ideas.

When we transfer the content onto a structural metamap, we can see where a DSRP analysis can predict the deeper questions that need to be asked to better understand the problem and determine a course of action.

[4] Waskom, R., et al. (2014). *U.S. Perspective on the Water-Energy-Food Nexus.* (Information Series No. 116), 6. Adapted from: Hoff, H., Understanding the Nexus. Background Paper for the Bonn 2011 Conference: The Water, Energy and Food Security Nexus. 2011, Stockholm Environment Institute: Stockholm.

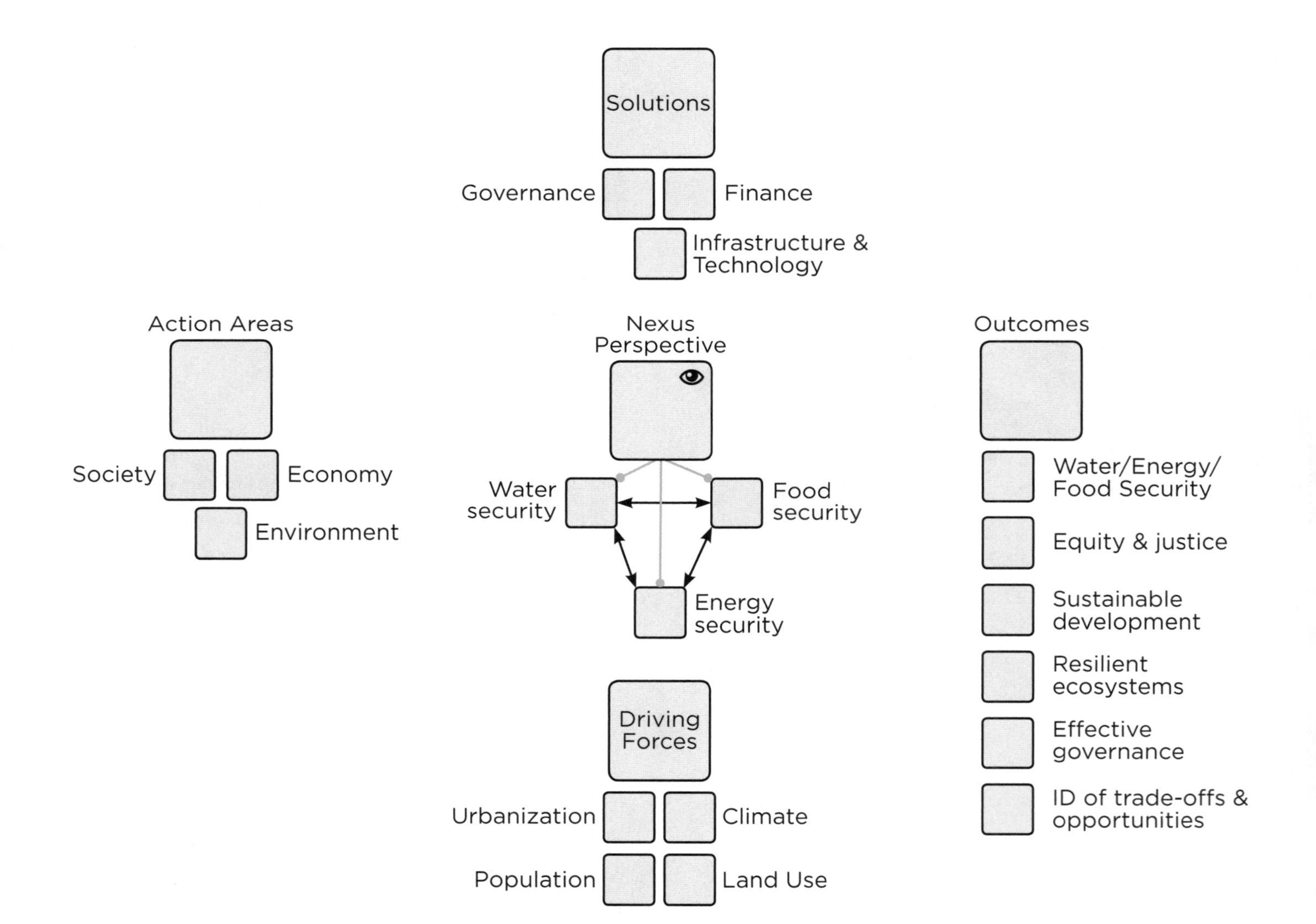

Figure 6.20: MetaMap of Nexus

First, we can start by recognizing the many distinctions being used, use DSRP to deconstruct them, and offer a common understanding of each of the terms used in the map. Specifically, the term "security" needs to be clearly defined to ensure all stakeholders in the solution process are working in concert with one another based on common understanding of the goal of water, food and energy security. Collective efforts could either miss the mark or make little sense if these key distinctions are not distinguished fully. Figure 6.21 depicts the structural map that tells us to ask the important question, "what do we mean by security? What is security comprised of?"

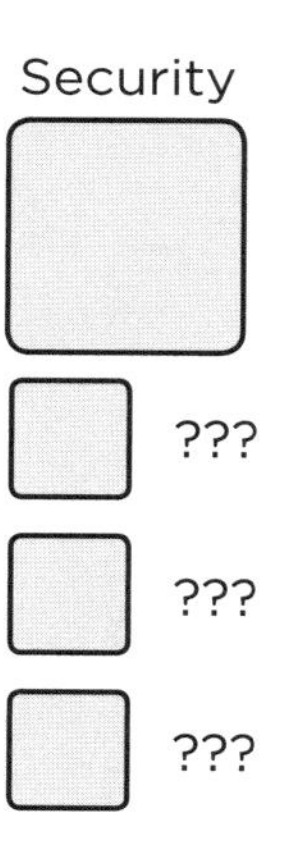

Figure 6.21: Distinguishing Key Concepts in a Map

Note also that the main idea of this research is the water-food-energy nexus, which is viewed as the crux of finding a solution to pending global resource shortages. A metamap depicting this idea structurally tells us that the nexus is comprised of three parts—water security, food security, and energy security—that are inextricably related. The thrust of the work that must be done is to thoroughly examine the critical relationships among these three parts, which is not depicted in the original graphic in 6.19. Contrast that to Figure 6.22, the map of the nexus perspective, which clearly (and predictively) shows that the crux of our future work is in the full exploration and examination of the relationships between and among the three security areas.

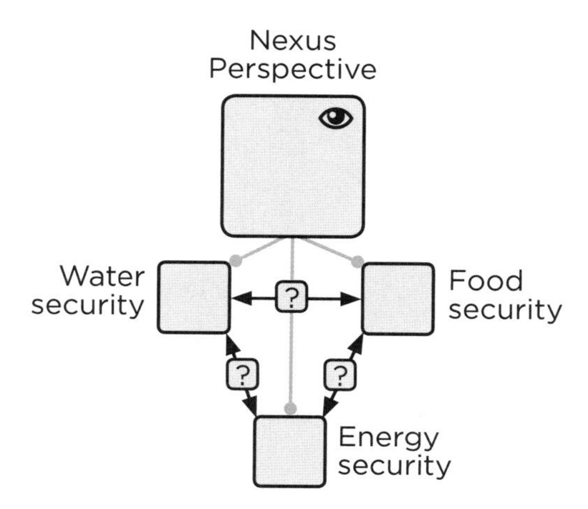

Figure 6.22: Structural Map of the Nexus Perspective

The originating graphic in Figure 6.19 is a depiction of the ideas generated from an important conference of educators, scientists, and policy makers who are hoping to address an impending resource crisis. What DSRP tells us is to examine the underlying structure of those ideas to better understand them, which will enable us to determine a viable course of action to take. Figure 6.23 rebuilds the basic map components in a way that better explicates what the interrelationships between Action Areas, Solutions, Factors, and Outcomes are in order to better represent what we do and do not understand about the problem in the first place.

In other words, Figure 6.23 lays out the main ideas in Figure 6.19 structurally, showing the critical relationships among the ideas, the distinctions that warrant further explication, and the crux of the work: the salient relationships among water, food, and energy security that require robust research to ameliorate the impending global resource crisis. DSRP predicts both the gaps we need to fill in our knowledge, and where best to spend our efforts to gain a deeper understanding and solution.

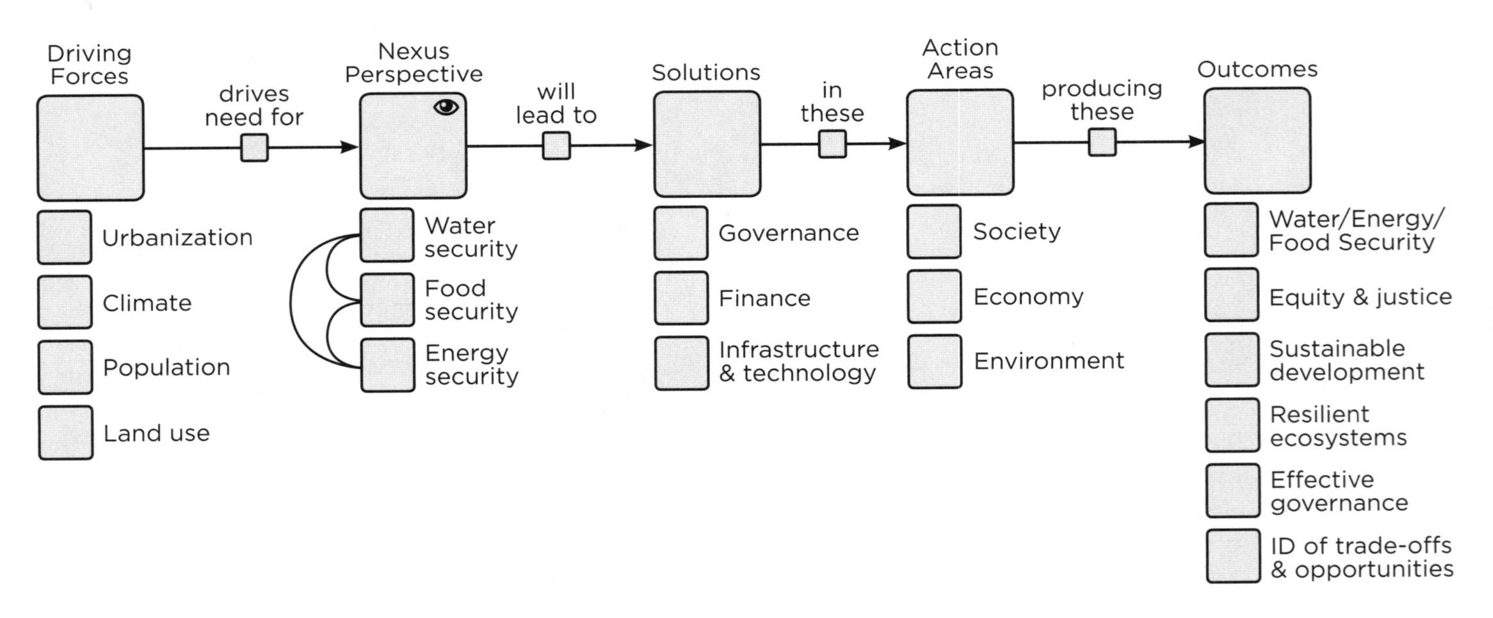

Figure 6.23: Information and Structure Map of the Nexus Research

CHAPTER 7

EMBRACE AND/BOTH LOGIC

A NEW KIND OF LOGIC

There is always an underlying logic implicit in both informal and formal systems thinking methods. Making the logic explicit leads to clarity of thought and deeper understanding of concepts. Logic is any system of principles that guides one's thinking. It need not be formalized or even conscious. All of us use logic every day without an awareness of what it is or where it came from. Systems thinking as a method also has an underlying logic.

The flaws of bivalent logic are pronounced as things get more complex, problems more wicked, and alignment with mental models more important.

BI-VALENT VS. MULTI-VALENT LOGICS

Bivalent logic takes two values. The word comes from the combination of *bi-* (two) and *-valent* (being strong). So we can think of bivalent logic as having two strong positions: right or wrong, on or off, guilty or not guilty, true or false, 1 or 0. These are all forms of bivalent logic invented by Aristotle. This and another of Aristotle's favorite logics, categories, is the basis for Western civilization.

It would be hard to overestimate the enormous impact that bivalent logic has had on society, both positive and negative. Bivalent logic is quick, easy, and clean. Unfortunately, in reality, nearly infinite shades of gray exist in between the logic of right and wrong, black and white, or even liberal and conservative. Bivalent logics lead to static and permanent ideas such as categories. You can't sometimes belong to a category and sometimes not. In general, bivalent logic is like law and order, it's nice to have just the right amount of it, but too much is oppressive, controlling, and curbs liberty. Bivalent logic is also *not* how the real world works, which can be a significant downside when accuracy is important. That's not to say that bivalent logic can't approximate the real world, sometimes remarkably well, but the flaws of bivalent logic are pronounced as things get more complex, problems more wicked, and alignment with mental models more important.

Multivalent logic has more than two outcomes. It can be difficult, messy, and slow. It can take more time to consider the multiple options. Worst of all, you won't fit in, because most people adhere pretty exclusively to the bivalent logic of their training. In contrast, if you're in an environment where questioning assumptions and biases or thinking creatively or systemically about things is encouraged, then multivalent logic is a good route to take. But wait a minute! Isn't contrasting these two forms of logic bivalent? Isn't there a middle way?

THE POWER OF AND/BOTH LOGIC

Maybe we "and/both" over "either-or." DSRP logic can be

bivalent and multivalent? Let's take a look at how it works. Let's choose the letter "A". Let's think of A as a variable for Anything. A can be anything you want it to be, including nothing at all. If you think of the letter A, you cannot think of it without all four DSRP patterns. Here's why:

If you think the idea A, then you must also think of the idea "not-A" because A cannot exist devoid of context. Of course, this may be happening unconsciously. Let's call A the thing and not-A the other. So far, so good—you've made a *distinction* between A and not-A.

Figure 7.1: Any identity (A) Implies the Other (Not-A)

A and not-A must be related because one literally implies the other.

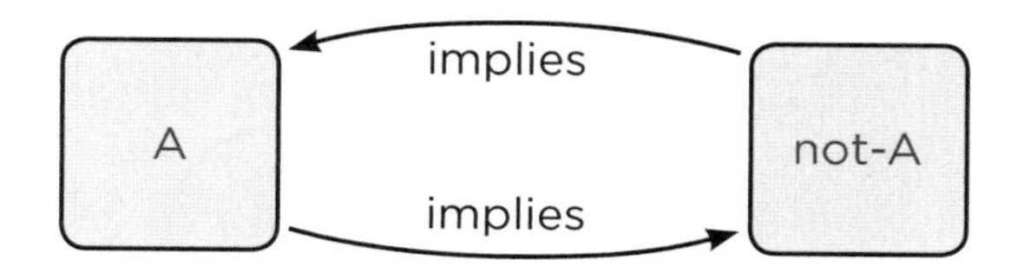

Figure 7.2: A Relationship of Co-Implication Is Necessary

So far, we can see that even for a thing as simple as the letter A, we require two structures of thought: a *distinction* and a *relationship*. What else is implied?

The simplest kind of system can be defined as a relationship between a whole and its part or parts. A and Not-A are two parts related to one another, and thus comprise a *system*. Now we see that for the simple thought A to occur, we need to be able to create a thing-other distinction, a relationship, and a part-whole system.

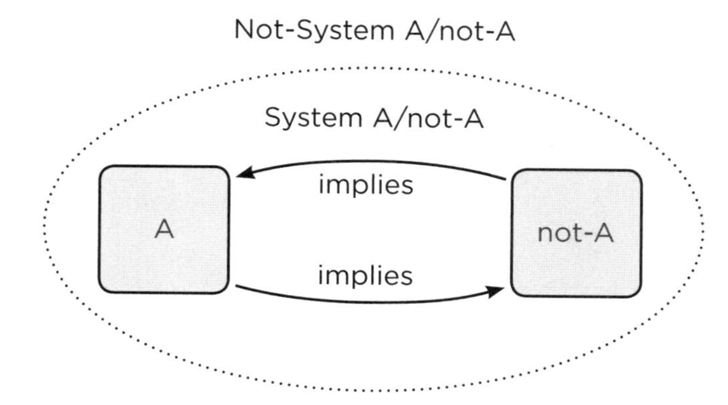

Figure 7.3: When Any Two Things Are Related, a System Is Formed (and a Not-System)

Notice, too, that the System A/not-A instantly implies a Not-System A/not-A. This is because as soon as you draw that circle around the system, you're creating a new boundary or distinction: the system and the not-system.

Now, notice something interesting. Not-A was defined by A in the first place. It's as if it didn't have an identity separate from A. But the truth is that the other stuff that is not-A has its own identity independent of A. It's just that because we're focused on A, we've defined this other stuff in terms of A.

This whole idea of not-A is based on a *perspective* because not-A is *defined in terms of A*. So in the same way that we started with A, we could have just as easily started with not-A. You can imagine that not-A might have its own identity, such as "B." We can see that even a simple idea like A or B requires a fourth element: *perspective*. Every perspective is made up of a point and a view.

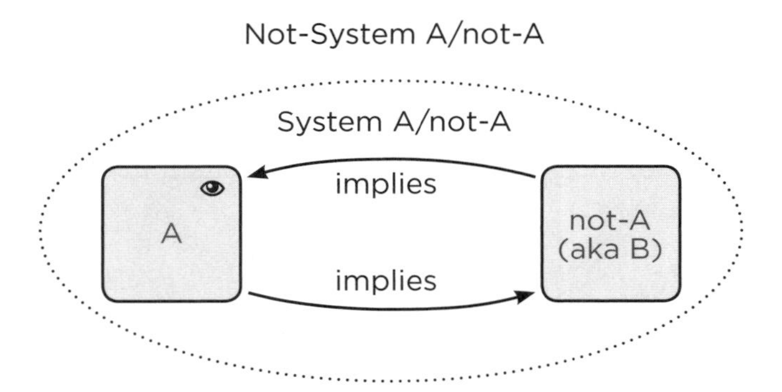

Figure 7.4: Perspective Is Also Necessary, B in Terms of A, or A in Terms of B

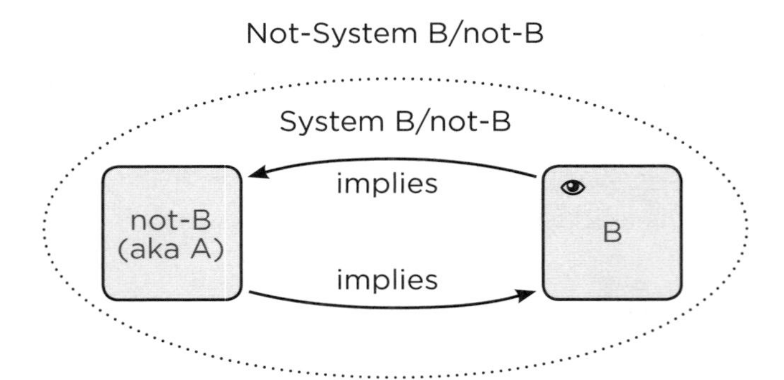

Figure 7.4 (Continued): Perspective Is Also Necessary, B in Terms of A, or A in Terms of B

If you think this is just an academic point, consider how perspective logic is used to recast one's political opponents in a negative light. Pro-life is recast as anti-feminist or anti-choice. Pro-choice is recast as pro-abortion or anti-life. There's a reason why we don't let our political opponents name us in terms of their position.

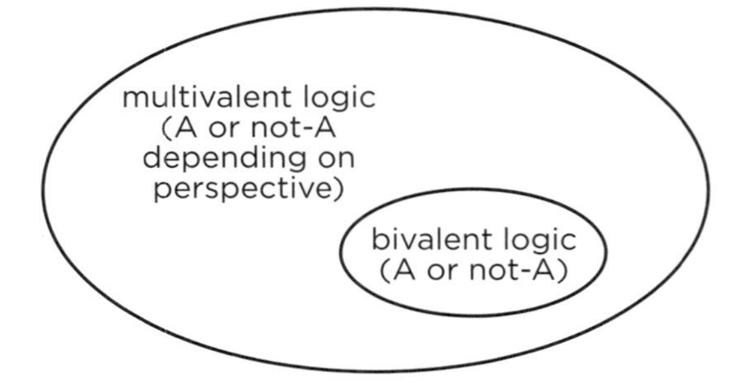

Figure 7.5: Bivalency Inside of Multivalency

There you have it. In order to have any idea or thing, which we called A, you need to make a bivalent distinction between it and not-it. In doing so, you've made a relationship, part-whole systems, and taken a perspective. This means we start with a *bivalent* system (A/not-A) and then realize that our bivalent system is sitting inside a multivalent system. It means that A/not-A only has meaning from the perspective of A and that all of the other possible distinctions one could make also have perspective. Therefore, there are infinite shades of gray in between A and not-A, unless you ask from the perspective of A, and then there's two. Bivalent logic is encapsulated in a multivalent system of logic.

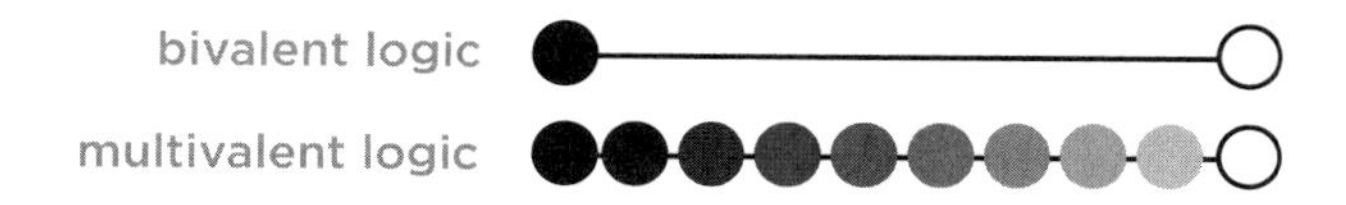

Figure 7.6: Old Vs. New Logic

P := $(e_1 \leftrightarrow e_2)$	A Pattern is defined as element$_1$ co-implying an element$_2$		
An element$_1$ exists	An element$_2$ exists	element$_1$ co-implies element$_2$	A Pattern exists
e_1	e_2	$e_1 \leftrightarrow e_2$	:= P

D := $(i \leftrightarrow o)$	A Distinction is defined as identity co-implying an other		
An identity exists	An other exists	identity co-implies other	A Distinction exists
i	o	$i \leftrightarrow o$	:= D

S := $(p \leftrightarrow w)$	A System is defined a whole co-implying a part(s)		
A whole exists	A part exists	whole co-implies part	A System exists
w	p	$w \leftrightarrow p$	:= S

R := $(a \leftrightarrow r)$	A Relationship is defined as an action co-implying a reaction		
An action exists	A reaction exists	action co-implies reaction	A Relationship exists
a	r	$a \leftrightarrow r$	:= R

P := $(\rho \leftrightarrow v)$	A Perspective is defined as a point co-implying a view		
A point exists	A view exists	point co-implies view	A Perspective exists
ρ	v	$\rho \leftrightarrow v$	:= P

$T(D \vee S \vee R \vee P) \rightarrow (D \wedge S \wedge R \wedge P)$		The existence of D or S or R or P implies the existence of D and S and R and P
If D or S or R or P exists	Then D and S and R and P exists	One Pattern implies all the Patterns
$(D \vee S \vee R \vee P)$	$(D \wedge S \wedge R \wedge P)$	$T(D \vee S \vee R \vee P) \rightarrow (D \wedge S \wedge R \wedge P)$

Table 7.1: Underlying Multivalent Logic of DSRP

A NEW KIND OF INQUIRY

Aristotle was of course Plato's best student and Plato was a student of Socrates. Socrates developed the Socratic Method of questioning which is still alive and well today, unlike Socrates himself. He was accused of corrupting young minds and subsequently was put to death (alternatively, perhaps because his Socratic Method of questioning was extremely irritating).

In its current day incarnation, the Socratic method (questioning, inquiry, etc.) is a popular way of getting students to think critically. The teacher focuses is on questions more than answers and on getting the students to engage their minds to try to answer these questions. It is an ancient but effective thinking and teaching tool. Its modern day utility after all these centuries is testimony to its value. But all things can be improved. The problem with Socratic questioning isn't the approach. It's the questions. The main thrust of the questions asked are born of the old 5 Ws: Who? What? Where? When? Why? How? Of course these questions will never go out of style. They are useful questioning tools. But there are questions, based on our multivalent DSRP logic, that allow for more robust and sophisticated answers. They are questions that can be used to explore any topic, system, or wicked problem. Like the 5Ws, not all the questions are appropriate all the time, but they are excellent prompts and have analogous metacognitive mappings. Each of these questions can be mapped in the form of a question.

The following "MadLib" style DSRP questions can be helpful in using the structure of DSRP to generate new ideas.

"MadLib" style DSRP questions.[1]

Distinctions

What is ______?
What is not- ______?
How would you distinguish between ______ and ______?
Can you compare and contrast ______ and ______?

Systems

What are the parts of ______?
What is ______ a part of?
Can you name some of the parts of the parts of ______ ?
What are the parts of the relationship between ______ and ______?
What are the parts of ______ when looked at from the viewpoint of ______?

[1] To work with these questions dynamically in a program that can visually map the question you choose, go to: cabreraresearch.org/thinkquery

Relationships

What ideas are related to ______ and what ideas are related by ______?

What idea relates ______ and ______?

How are the parts of ______ related?

How are the parts of ______ related to the parts of ______?

What are the relationships among ______ and ______, and other things?

Perspectives

What are the parts of the viewpoint ______ when looking at______?

How are ______ and ______ related when looking at them from a new perspective?

Can you think of ______ from multiple perspectives?

What are the parts of ______ when looked at from multiple viewpoints?

D + S + R + P

Tell me all you know about ______ by DSRP'ing it.

CHAPTER 8

EVERYDAY AND ADVANCED APPLICATIONS OF DSRP

THE IMPORTANCE OF UNIFYING THEORIES

As a universal theory of systems thinking, DSRP grounds newcomers and old-timers alike and furthermore grounds the field. This has been the case in other fields. For example, before the fundamental theories of Newton, the field that we know of today as physics was an immature meta-physics. Before the fundamental theories of Antoine Lavoisier, the mature scientific field that we know of today as chemistry was a magical alchemy. Before the fundamental theory of Darwin, the sophisticated field we know of today as biology was little more than stamp collecting (the identification and categorization of species). The great geneticist and evolutionary biologist Theodosius Dobzhansky wrote, "nothing in biology makes sense except in the light of evolution." DSRP provides a fundamental theory of systems thinking. With it, everything in systems thinking makes sense.

A shared language based on DSRP increases cross-pollination among methods and approaches and will lead to co-evolution of all methods, because we can see their relative utility across domains and problems. This will facilitate specialization within a powerful, unified systems thinking community.

When we examine the MFS Universe, we see that all of the systems thinking methods are using DSRP rules. Some systems thinking corresponds to some rules more than others, while some adds more specific detail to the simple rules (e.g., specifically highlighting feedback-type rather than more general relationships). This means we can enjoy the plurality of MFS systems thinking while appreciating its universal simple DSRP rules. This has some important implications for systems thinking. It leads to less cannibalism and fewer silos within the field. It is often the case that for a systems thinking method to succeed its proponents must disparage the value of other systems thinking methods (i.e., cannibalize) or isolate the method as superior (i.e., silos). Universal rules like DSRP elucidate that all systems thinking methods come from the same philosophical DNA and are merely useful adaptations to tackle certain types of problems. A shared language based on DSRP increases cross-pollination among methods and approaches and will lead to co-evolution of all methods, because we can see their relative utility across domains and problems. This will facilitate specialization within a powerful, unified systems thinking community. DSRP rules offer:

- A common language, grammar, and structure for all systems thinking and systems sciences;
- Insight into where various specialized methods can improve and where they stand out (e.g., which methods best fit certain types of problems or applications);
- Co-evolving generalization and specialization func-

tions in the field, which increase adaptation, cross-pollination, robustness, and diversity, while decreasing silo-fication and tribalism;
- Accessibility, which increases the field's hospitality to newcomers and therefore its adoption rate by decreasing frustration and drop off and increasing the ease with which systems thinking can be deeply understood and used; and
- A new and definitive systems logic for developing new methods and approaching wicked problems.

As we dive deeper into specific systems thinking applications, it helps to break them into some useful part-whole groupings to give us clarity. Let's consider a few groups:

1. *Everyday tools that are applied using a systems thinking lens:* These are abundant, often times more so than formal methods. Buzanian mind maps, Novakian concept maps, Ishakawa (or fish-bone) diagrams, brainstorming splat maps, hierarchical and process diagrams, and tables, graphs, and outlines are used ubiquitously in systems thinking. A review of their structure can guide us to their better use.
2. *"Systems _____:"* approaches where a patchwork of both formal and informal methods and ideas of systems thinking is applied to a discipline or field (e.g., systems evaluation, systems engineering, systems biology, and systems neuroscience). This changes what or how a discipline thinks about itself.
3. *Formal systems thinking methods:* Usually developed for specialized and often technical and advanced applications, these formal methods make up the MFS Universe. We will explore a diverse sample of the most popular ones such as network theory, system dynamics, and soft systems methodology to see how an understanding of DSRP can bring more sophistication to these methods.

In what follows, we'll take a look at examples in each of these groups to give you a better sense of how the universal patterns of DSRP can make you better at applying any of these methods.

EVERYDAY TOOLS

While there are many formal systems thinking methods like system dynamics, soft systems methodology, and network theory, the truth is that much of the everyday systems thinking that occurs is applying everyday tools. These tools will be familiar to most of us: tables, graphs, flow charts, outlining, tree diagrams, and brainstorming. In practice most systems thinking isn't quite as formalized, technical, or advanced as the methods we will review later in this chapter. Most systems thinking is done rather *ad hoc*, piecing together any number of systems thinking concepts like a patchwork quilt to be applied to a particular wicked problem or everyday task. In so doing, this patchwork includes a number of tools and techniques that wouldn't be considered systems thinking, per se. For example, most people wouldn't think of a common graph or data table as an artifact of systems thinking. What we hope you will realize is that these valuable tools take on even more power in the hands of a systems thinker and their utility can be extended with DSRP.

XY GRAPHS

Graphs are immensely useful in every discipline of knowledge. Understanding DSRP not only reveals the implicit structure of graphs but also allows us to extend graphs and to make them better. Figure 8.1 illustrates that every XY graph is a relationship between two systems (the system of X and the system of Y).

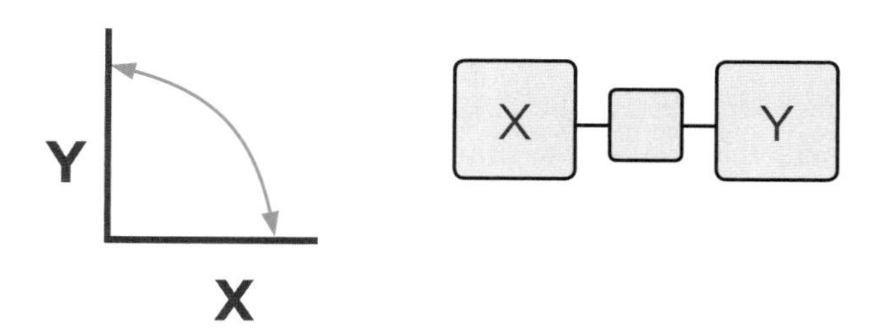

Figure 8.1: X and Y Axes Are Both Distinctions in a Relationship

Figure 8.2 shows that X and Y are also systems made up of parts (e.g., the numbers on their axis), each of which are individual distinctions.

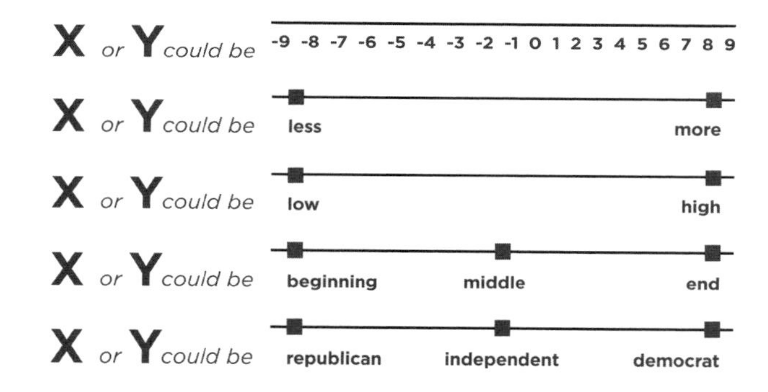

Figure 8.2: Each Axis is a System Made up of Distinctions

By combining the points illustrated in Figures 8.1 and 8.2, Figure 8.3 shows something surprising: every graph is actually an R-Channel (see Chapter 5). R-Channels are common cognitive jigs where the parts of two systems are being related. In addition, all graphs also have a root perspective and many graphs could benefit from using various perspectives to analyze them further.

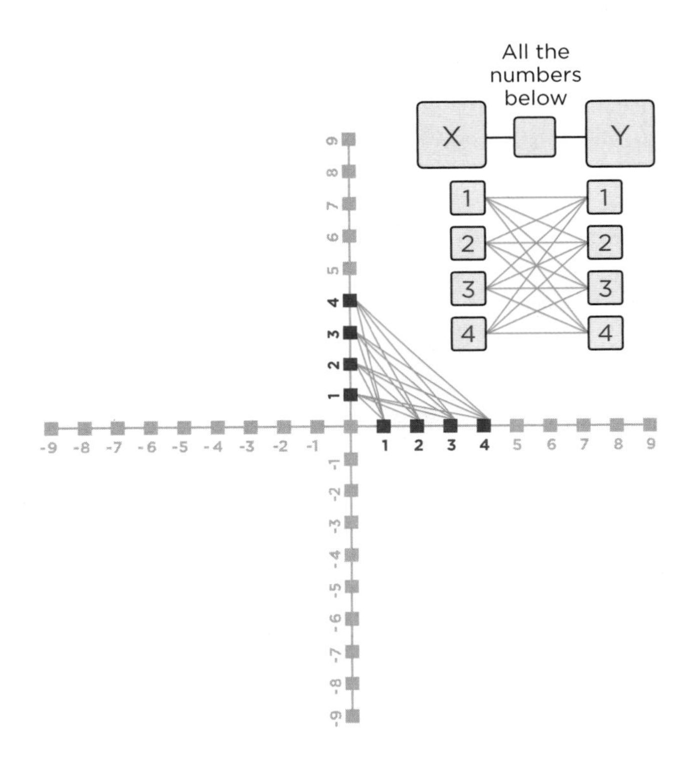

Figure 8.3: Every Graph Is an R-Channel

TABLES

Tables are another common organizational tool. From the periodic table of elements to a recipe for ganache, from kindergarten classroom routines to doctoral appendices, the time-tested, tried-and-true, trusted, and downright handy table is one of mankind's favorite meaning-making devices. You know of course that tables are made up of cells and the cells contain some type of data or information. These cells are arranged in columns (vertical cells) and rows (horizontal cells). The purpose of a table is to arrange the data in a way that gives it the meaning you are trying to communicate. The cells allow us to easily place data, and the arrangement of those cells in relation to one another is what gives the information meaning. Generally speaking, tables are great for quickly and easily laying things out that relate on two axes—the column and the row. But what does it all mean? Is there an underlying structure to tables that is fundamental? The answer is yes: DSRP.

The first thing to notice is pretty simple. Every cell is a distinction. That is, all the different pieces of data or information are distinctly different. Even if the content inside the cells is the same, the location of the cell makes them different.

Next we can see that part-whole systems are implicit in table structure, because the columns and rows are natural groupings. The metamap in Figure 8.4 shows that merged columns, columns, and rows are just systems made up of parts. The metamap elucidates this structure using nestedness and size.

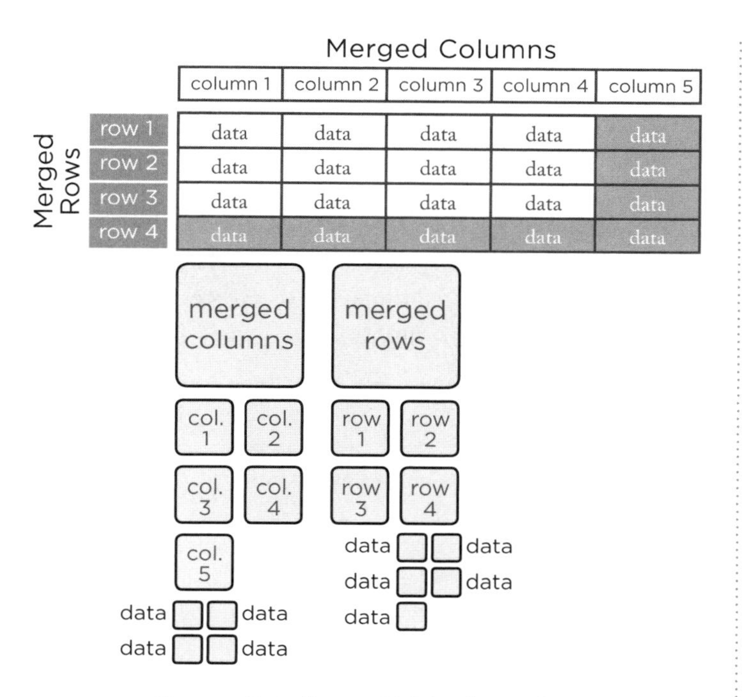

Figure 8.4: Every Table Contains Part-Whole Systems

When we zoom in to individual cells in columns and rows, we see that like a graph, there are myriad Barbell-relationships. The data that lies at the intersection between a row and a column is actually the relationship between them. For example (in Figure 8.5), column3 and row2 are related by the data in cell 3, 2. All of the columns and rows are relational in this way, so we can think of a table as an elegant way of making lots and lots of barbell relationships.

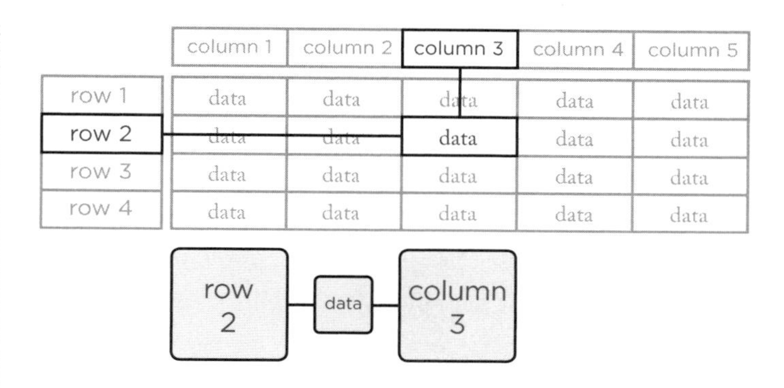

Figure 8.5: Tables Are Full of Barbell Relationships

We can also use tables to create easy to understand P-Circles which we call P-Circle Tables where the items in one header column or row are looking at the items in another header column or row as in Figure 8.6.

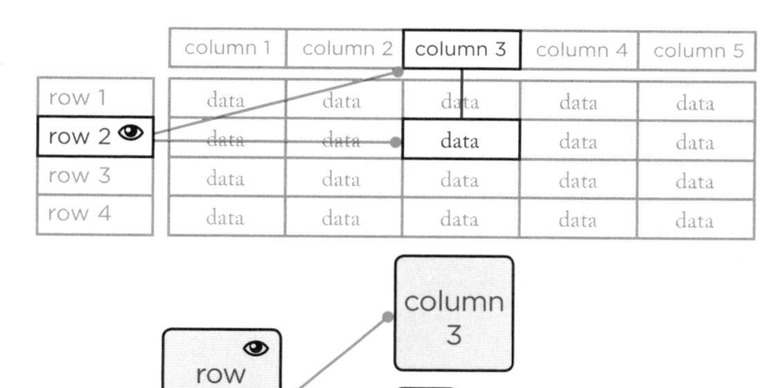

Figure 8.6: A P-Circle as a Table

A simple example of what perspective tables look like is helpful. Table 8.1 illustrates a 5th grade Early Hominids project using a P-Circle in table form. This table shows early homonids (e.g., Australopithecus, Homo *habilis*, Homo *erectus*, Neanderthal, Homo *sapiens sapiens*, and Cro magnon) being looked at from multiple perspectives (e.g., tools, diet, behavior, looks, fossils). For example, when we look at Australopithecus from the perspective of tools we can see that they had razor sharp rocks, but when we look at Homo *habilis* from the same perspective, we see that they were making tools from other tools.

BRAINSTORMING OR "SPLAT MAPS"

Despite meta-analysis research that indicates that brainstorming doesn't work,[1] and although it is not systems thinking *per se*, brainstorming remains a popular technique for group and individual systems thinking. We call the results of brainstorming a splat map. Splat maps are not only created through a group process, many individuals are capable of creating these splat maps on their own. Mind Map is a popular tool for cre-

1 Mullen, B., Johnson, C., & Salas, E. (n.d.). Productivity Loss in Brainstorming Groups: A Meta-Analytic Integration. Basic and Applied Social Psychology, 3-23.

Table 8.1: A P-Circle Table of Early Hominids From Multiple Perspectives

	Austalopithicus	Homo habilis	Homo erectus	Neanderthals	Homo sapiens sapiens	Cro Magnon
From the Perspective of Tools	razor-sharp rocks	made tools from other tools	spears, torches, axes	spears, torches,	spears, torches,	spears, torches, axes
From the Perspective of Diet	meat, fruit, roots, insects	meat, fruit, eggs, insects	meat	vegetables, meat	vegetables, meat	vegetables, meat
From the Perspective of Behavior	Africa, groups, hunters	Africa, groups, hunters	campsites, groups,clothes, cooked	Europe, funerals, clothes	hunter gatherers	Europe, jewelry, clothes
From the Perspective of Looks	3 to 5 feet, 80-113 lbs, short, stocky	strong, round skulls, thin	5'4" tall, slender, 140 lbs	tall, slender, 4-5 ft tall	biggest brains	big forehead
From the Perspective of Fossils	"Lucy"	Tanzania, Jonathan Leakey	France, skulls, jaws, teeth	none	none	none

ating splat maps because it gives the illusion of meaningful structure to otherwise unstructured ideas (i.e., a splat map). Post-it notes are also a popular technique for brainstorming and splat maps.

It's fine to create splat maps but once you're done with the brainstorm you need to do a little cleanup to get some meaning out of it all. This requires DSRP organization and structure.

Start with all of the notes or ideas that have come out of the brainstorm and begin processing them more thoroughly in the following steps (not necessarily in order):

- Ensure that each note or idea is distinctly different; combine notes or ideas that are the same but perhaps use different terminology;
- Begin organizing into part-whole systems things that belong together; pay attention to see if some of the notes or ideas are a whole that contain some other notes as parts or look for a bunch of notes or ideas that are parts and create a new note to represent the whole they belong to;
- Look for relationships between and among the ideas or notes; try to add a new note to represent what that relationship is; and
- Identify some of the perspectives that are in play that lead to the distinctions being made; perhaps some of the notes or ideas are perspectival as well.

The basic idea is to go from something that looks like the board in Figure 8.7 to a mental model that has more structure, more meaning, and can be readily understood.

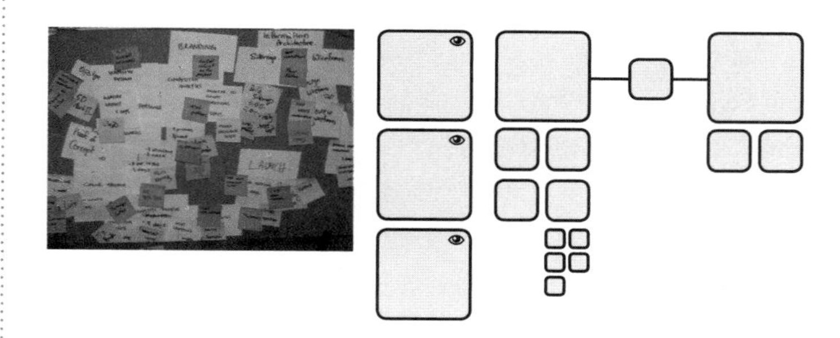

Figure 8.7: Use DSRP to Structure Your Splats

PROCESS MAPS

Another popular style of maps used in systems thinking are process maps or flow charts that use spatial layout and relationships to reflect the order of things or ideas. Figure 8.8 shows a few examples of process maps.

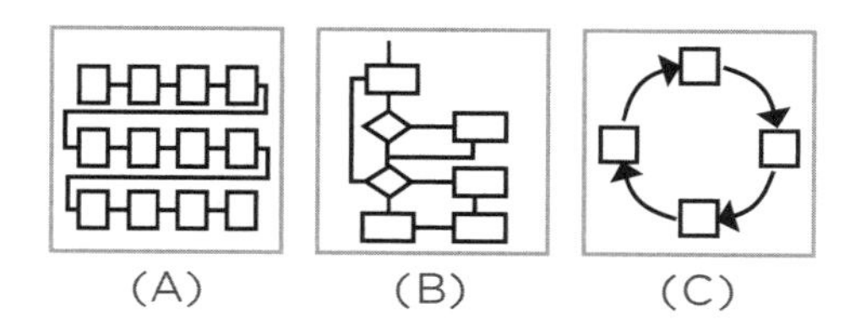

Figure 8.8: Examples of Process Flow Maps

It is easy to see that metamaps and DSRP not only provide the foundation for such maps, but can also be used to extend them. Items in the process can be better distinguished or broken down into parts. Relationships in these diagrams can be further explicated using RDS-Barbells. And perspective can be used to further elucidate basal assumptions or to consider alternatives.

ISHIKAWA OR FISHBONE DIAGRAMS

Relational maps such as Ishikawa or "fishbone" diagrams (See Figure 8.9) are maps with repeating part-whole and relational structures. In an Ishikawa diagram, each part is causally related to its whole in a repeating chain. The purpose of Ishikawa diagrams is to show all the nested causes that lead to an effect.

Based on DSRP, MetaMap can easily model these structures (See Figure 8.10). The same Ishikawa diagram in MetaMap shows the distinctions, part-whole relationships and fractal dynamics more clearly. MetaMap can extend systems thinking by detailing the relationships, further deconstructing the systems, and adding perspectives. We can see that the structure flows from small part-causes to larger ones, culminating in a whole effect.

We can see that a fishbone map is a simple repeatable pattern: parts pointing causally toward their wholes. We also clearly see that we are dealing with a single system made up of parts, which is not well communicated in the original fishbone diagram (Figure 8.9), as the endpoint label on the right actually contains all the other elements and relationships in the map.

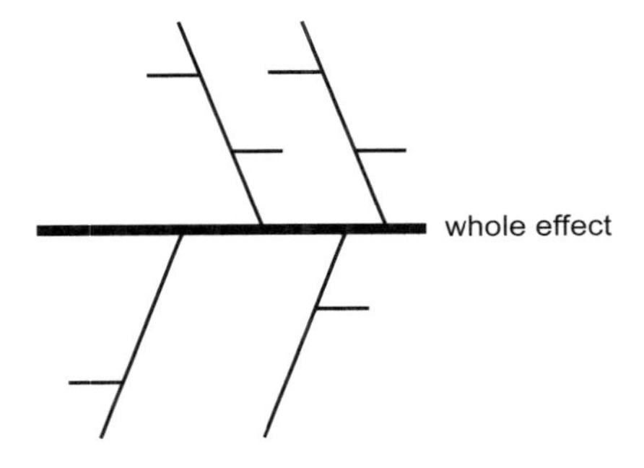

Figure 8.9: Ishikawa Diagram

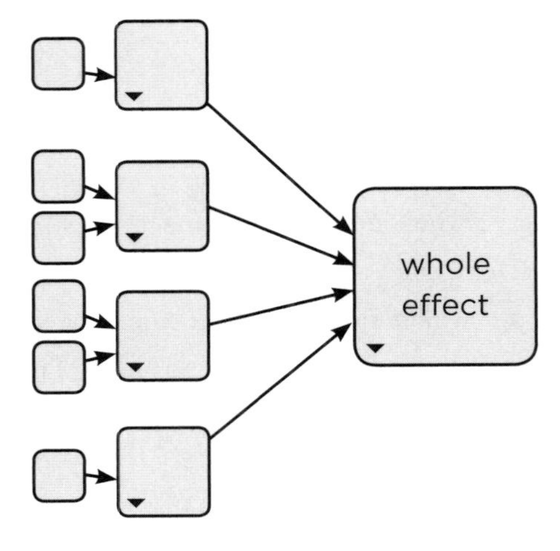

Figure 8.10: Ishikawa Diagram in MetaMap

NOVAKIAN CONCEPT MAPS

A popular type of map, called a Novakian Concept Map, elucidates an important change that needs to be made in our visual grammar. As you can see in Figure 8.11, relationships are identified with words or phrases, which is a good thing. However, the types of words or phrases used means that relationships are given a lower status than the things they relate. Whereas the ideas of things being related can be any possible concept, the words or phrases used to describe relationships are of a particular ilk, chosen from a finite set of possible words/concepts such as: *leads to, is part of, becomes, causes, is,* and *contains.* In Concept Maps, word labels are related by lines with arrows. The lines indicating relationships can be left blank or receive a "linking word" as a label.

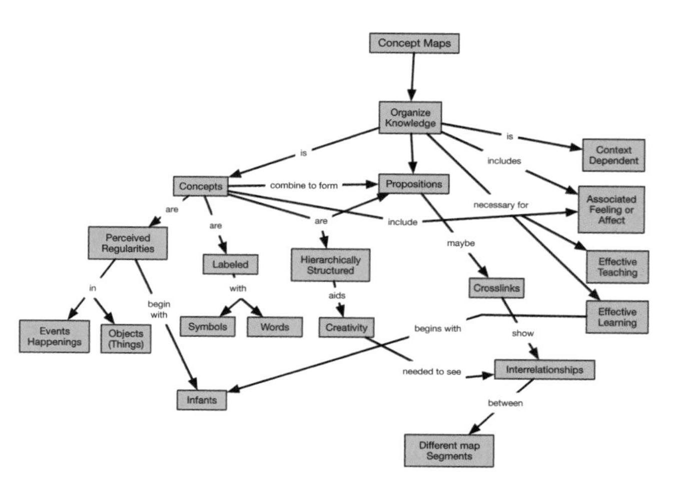

Figure 8.11: Novakian Concept Map

Both in design and in practice, however, the relational lines do not receive the same treatment as the things they relate. The things being related are thought of as structural and get a border around the word, whereas relationships are lines with words.

The concepts that are attached to relationships are no different than the elements being related.

Metamaps and DSRP suggest that relational ideas or things can be as vast and infinite as any of the things they relate. That is, the ideas of things that can be attached to a relationship are as vast and varied as the ideas of things that exist. Yet we know that relationships between and among things in the world can be quite large and are not always "verb-like," "action," or "linking" words. The relationship between two things could be any concept or thing. For example, *marriage* might be the relationship between two people. A *US diplomat* might be the relational intermediary between Palestinian and Israeli negotiators. A doctoral dissertation might focus on *a whole web of complex causes and factors* that form the relationship between the dependent and independent variables. The relationship between a person's motivations and behavior might be a *mental model.* It's clear that where systemic thinking is concerned, the possible labels and concepts that might be attached to a relationship are not a finite set, but are actually infinite. *Any* thing or concept might form the relationship between two things.

Likewise, the elements that are being related could be an infinite set. We could think deeply, for example, about marriage, distinguishing various elements of marriage and relating these elements. Having elaborated marriage as a system, we could use that complex concept as a single relationship between two people. Therefore, relational ideas and things are different only because they "live on the line" that relates two ideas or things, not because of anything inherently different about the relational concept itself.

This means that all of the lines that are being used across various graphic organizers have enormous untapped potential. Figure 8.12 reminds us that any of the relational lines in a Novakian Concept Map could potentially be an RDS Barbell.

Thinking more deeply about, and zooming into, the relationships between and among things in order to see the vast worlds of complexity that often exist there is essential to systems thinking. Having a visual grammar that permits us to do so easily and consistently is important.

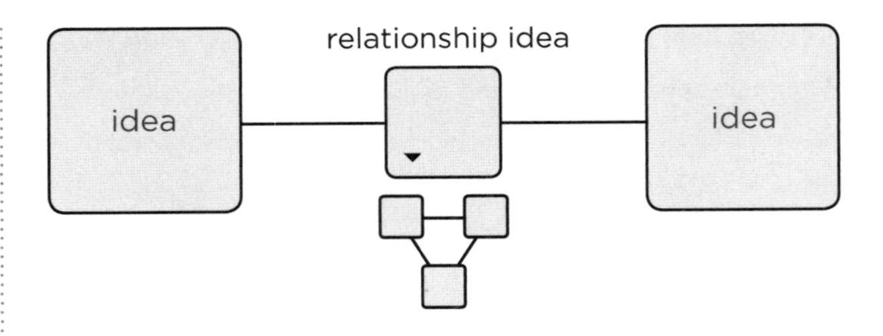

Figure 8.12: An RDS Structure Could Be Added to any Relationship in a Novakian Concept Map

Upon these lines can be bestowed tremendous meaning and clarity. Thinking more deeply about, and zooming into, the relationships between and among things in order to see the vast worlds of complexity that often exist there is essential to systems thinking. Having a visual grammar that permits us to do so easily and consistently is important. We can do anything to the relational line that we can do with any concept. We can distinguish it by adding a square. We can break the relationship into parts, and parts of parts. We can further relate the parts of the relationship. We can apply perspective on, from, or inside of the relationship.

HIERARCHICAL TREES

Hierarchical trees are pervasive and popular ways to organize information. All hierarchical trees are part-whole structures so when you see one in any context, just remember that the Systems Rule is at work. The problem is that when individuals use the Systems Rule exclusively (without relationships, distinctions, or perspectives), they inevitably end up in a thinking *cul de sac*. Part-whole hierarchies or trees are good to use, but they are not good enough.

One of the most popularized hierarchical trees is called a Mind Map. Even if you've never used a mind map, you have likely seen one and even if you haven't seen one, you've likely been influenced by one. Mind Maps aren't popular because they're good at anything, they're popular for the same reason that Cheetos are popular: not because they are packed with nutrients, but because they taste good. Mind Maps are easy to use and easy to feel successful with, but they lead to visual malnutrition.

To use Mind Maps is to love them. They are simple and seductive. They make us feel like we are getting somewhere, understanding things better, getting our thoughts on a page. But Mind Maps are based exclusively on a radial, centralized cognitive architecture that is not only scientifically invalid, but meta-cognitively destructive. The cognitive style of Mind Maps, like the cognitive style of PowerPoint,[2] causes thinking and communication biases. Mind Maps always result in a radial hierarchy even if that is not how the system being thought about is structured. Mind Maps make your ideas worse by forcing you to think inside of unnatural hierarchical and radial structures, which reinforces bad habits in systems thinking. They also increase map junk—unnecessary use of color and clip art that hides the overly simplistic structure of this technique.

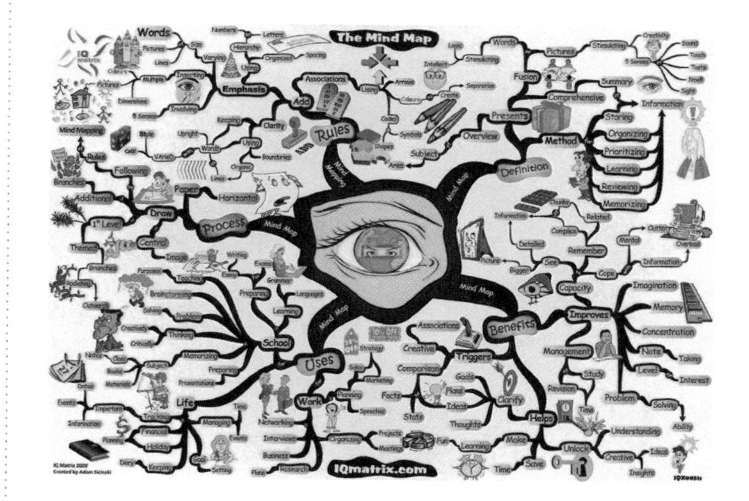

Figure 8.13: A Typical Buzanian Mind Map

Many people are fooled into thinking Mind Maps are creative, mind expanding, divergent thinking tools when in actuality they are structurally no different than organizational charts (org charts). Few people would argue that an org chart is the height of creativity, innovative thinking, or systems thinking. Yet notice that the two maps in Figure 8.14 are structurally the same. The diagram on the left is an org chart or hierarchical tree structure in four directions. Likewise the Mind Map on the right is a hierarchical tree in five direc-

[2] Tufte, E. (2006). *The Cognitive Style of PowerPoint: Pitching Out Corrupts Within.* Cheshire, Connecticut: Graphics Press.

tions (five black arms emanating out of the creepy eyeball). Both maps are structurally based on hierarchies radiating outward from a central point. How do the added color, clip art, and fonts increase understanding, creativity, divergent or systems thinking? The answer is, they don't.

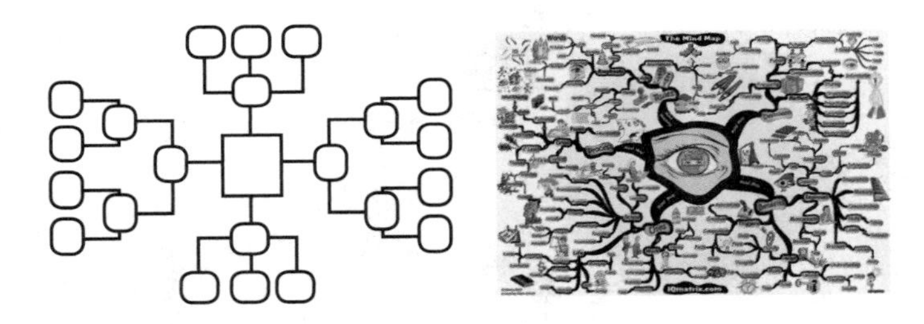

Figure 8.14: A Typical Buzanian Mind Map

The problem with Mind Maps is that the underlying "science" is based on three flawed ideas about your mind: (1) your mind thinks only hierarchically, (2) your mind thinks only in a "radial" way (i.e., all thinking "radiates" out from a single idea), and (3) the only cognitive structure needed to understand everything is part-whole. The truth is that your mind doesn't work this way. In contrast, metamaps utilize four universal patterns (DSRP), in the same way your mind was designed to think.

The part-whole structure of the Systems Rule creates hierarchies, so DSRP and metamaps support hierarchical thinking, but only in the context of testing boundary distinctions, interrelationships, and perspectives. Without these other rules, one's thinking becomes a forced march through one tree-like structure after another.

All of the part-whole structures (i.e., hierarchical trees) in Figure 8.15 are structurally the same pattern. This includes org charts (4), Mind Maps (1 and 2), bracket diagrams (5), many other styles of trees (3), and even outlines (6). All are characterized by part-whole structure (S) and all would benefit from additional analysis of the distinctions (D), interrelationships (R) and perspectives (P) involved.

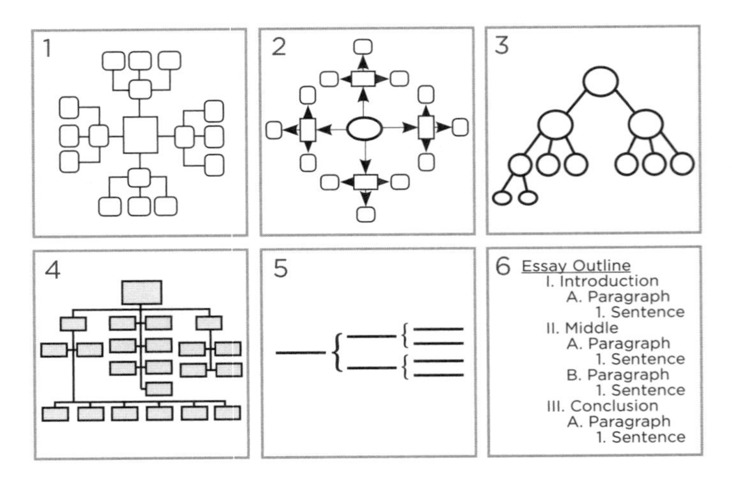

Figure 8.15: Hierarchical Tree Structures

One example of the benefit of adding D, R, and P to the hierarchies of S is in outlining (a task for which Mind Maps is often used in school). The outline has been used for centuries as a simple yet powerful tool for laying out a sequence of ideas in preparation to create a written work. The traditional outline uses Roman numerals to generate a simple, hierarchical part-whole structure. This same traditional approach is taught in classrooms and used by writers around the world. Yet teachers often complain that students fail to write in a way that relates one part of an essay to another. Clearly outlines may not be serving us well.

The solution to this problem is to help learners see both the underlying structure of writing and what's missing from a typical outline. A traditional outline has only part-whole structure and distinctions. That is, each item in the outline is (presumably) a distinct idea and the ideas in an outline are placed in hierarchical or part-whole structure.

When written works lack a coherent thread, it's because writers are putting a bunch of ideas into a part-whole structure but failing to think about the relationships that link them all together. This can be remedied by incorporating relationships. The simple rule is to relate each section to the one that follows using a part of the preceding section that we call a transition (see Figure 8.16). A transition relates one section of writing to the next. Depending on the scope of the writing project it could be a few words, a single sentence, a paragraph, or several pages. Both the introduction and the conclusion of written work should relate to each other and to the interior sections of the document.

To extend the power of traditional outlines even further, we can add perspectives (see Figure 8.17). Perspectives can be thought of as parts or sections of the document that provide "framing" for other parts—a lens through which the author is approaching other ideas. For example, in Figure 8.17, we have thought of the first section of the beginning, middle, and end as perspectives on the section. Here the author might lay down some of the assumptions, presumptions, precepts, or concepts that govern the discussion that follows, keying the reader into the point of view taken by the author.

By adding DSRP structure to traditional outlines, students focus not merely on the information that goes into each section, but on the structure of their written work (e.g., the relationships between sections) and how to better communicate their ideas to the reader.

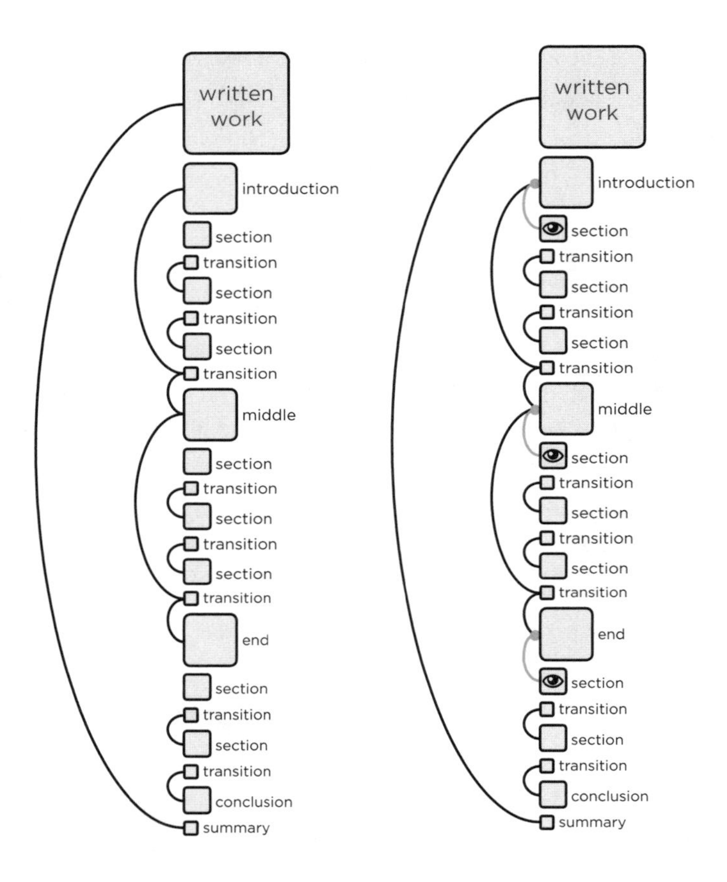

Figure 8.16: Outline with Relationships

Figure 8.17: Outline with Perspectives

"SYSTEMS ________"

By far the most popular application of systems thinking isn't a single formal method like system dynamics, soft systems methodology, or network theory (discussed later in this chapter), but a patchwork quilt of systems ideas applied to a particular field. In the past few decades, many scientists simply take systems ideas and apply them to create new sub-fields and disciplines of study. Systems engineering, systems biology, systems neuroscience, systems evaluation, and systems thinking in public health are all examples of this trend.

The new logic and inquiry methods of DSRP (discussed in Chapter 7) make it even easier to apply systems thinking to any discipline or method used within that discipline. We have come to call these approaches to systems thinking "Systems X" or "systems ________" (fill in the blank). What one learns is that most of these disciplinary approaches to systems thinking are simply applying systems thinking to a particular type of problem, context or discipline. So, for example, there is no "systems engineering" approach, *per se*. It is really taking a systems thinking approach to all the stuff engineers already do. Over time, these new systems thinking fields begin to form cultural norms around certain accepted ideas and practices.

This is perhaps where systems thinking may have the biggest impact in the sciences. Any one of the over 8,000 academic disciplines and 50,000 academic journals could be revolutionized by applying a systems thinking lens to their concepts, methods, and knowledge. If you are a young doctoral student, for example, and wish to make headway in your field, consider becoming an expert in systems thinking and begin applying its ideas and methods to your field. In the last decade, we have been involved in well-funded efforts to do exactly that in a handful of fields such as STEM science, public health, education, and evaluation. Each of these areas has tremendous potential for those interested in researching how to better apply systems thinking concepts to an existing field.

When we take a multivalent view, it becomes clear that universality and pluralism need not be mutually exclusive. The rich diversity of specialized methods is made whole by the universality of DSRP. Universality makes everything in systems thinking make sense.

FORMAL SYSTEMS THINKING METHODS

The universality of DSRP can transform the way we understand and use everyday tools like concept maps, graphs, and data tables. It can also improve how we add a systems thinking lens to existing disciplines. But can it really transform formal systems thinking methods designed for specialized and often advanced and technical purposes? Yes. Let's take a look at some of the most popular and promising methods of systems thinking and briefly look at how an understanding of the simple DSRP rules can transform these powerful approaches for the better.

SYSTEM DYNAMICS

System dynamics is a method that is popular, and for some types of systems—especially population models—system dynamics is very powerful. The basic idea behind system dynamics is a particular type of relationship called feedback. For the system dynamicist, feedback can be positive or negative (+/-) and there are two kinds of feedback loops: balancing and reinforcing.

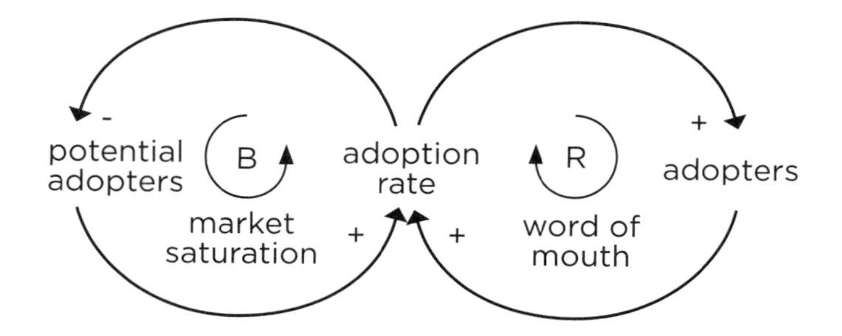

Figure 8.18: Reinforcing and Balancing Feedback Loops

The example in Figure 8.18 shows two feedback loops that are interacting with each other. A reinforcing loop called "word of mouth" shows that more adopters means a higher adoption rate, which in turn means more adopters. The other feedback loop is a balancing loop because when potential adopters increases, the adoption rate climbs, but as the adoption rate climbs, potential adopters decline (because those who have adopted are no longer on the market to adopt). This balancing loop is called "market saturation." Next, systems dynamics adds the concepts of "stocks and flows" (Figure 8.19). You can see, for example, that the feedback loops above led to accumulations of potential adopters becoming actual adopters based on the adoption rate.

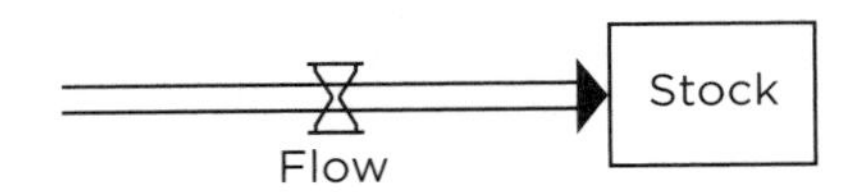

Figure 8.19: Stocks and Flows

We can think of these accumulations as a "stock" (something akin to water collecting in a bathtub) and the rates as "flows" (akin to the amount that the faucet is open to allow water into the bathtub).

When we combine these basic elements of system dynamics modeling (feedback loops and stocks and flows) we get more complex diagrams (Figure 8.20) that show multiple feedback loops and cycles, stocks, and flows interacting in a system.

One of the powerful things that system dynamics software allows you to do is attach equations to these conceptual drawings and then model the dynamics of the system so that you can simulate real-world conditions.

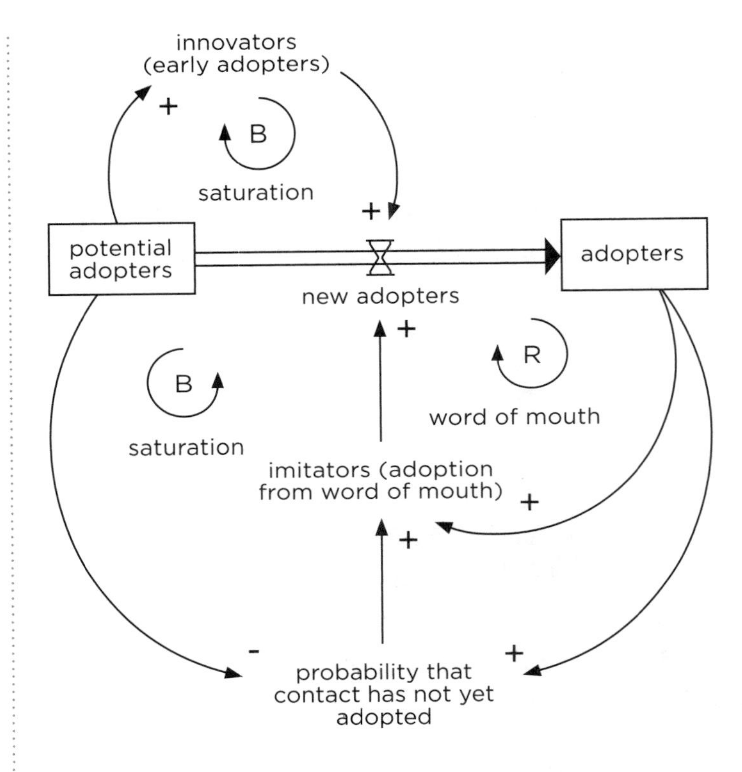

Figure 8.20: Complex Feedback Systems

Now let's take a look at what underlies system dynamics. First, we have some distinctions (Figure 8.21). In a system dynamics diagram, the distinctions are the words and symbols (in orange):

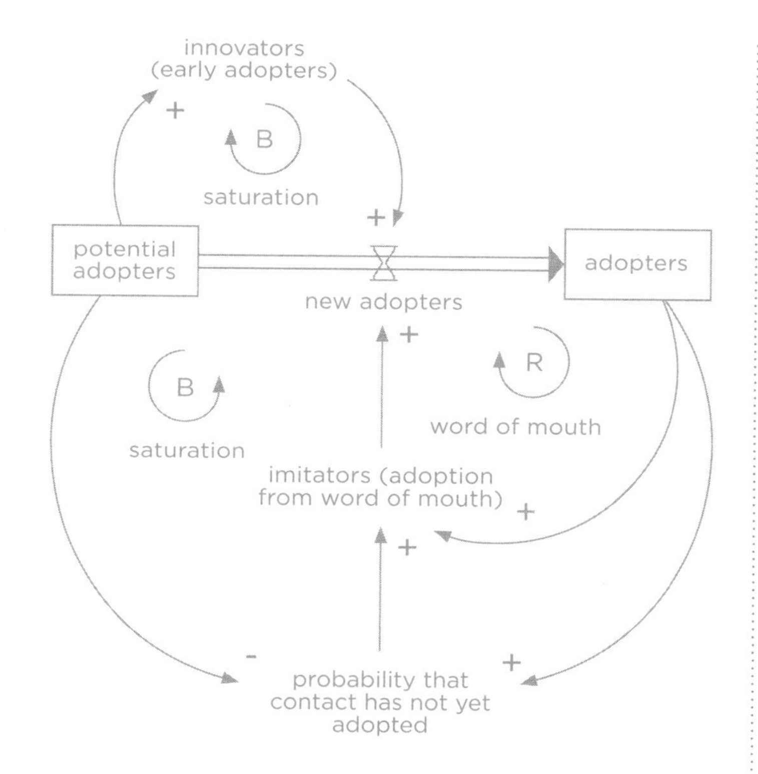

Figure 8.21: Distinctions in System Dynamics Diagrams

Figure 8.22 shows the relationships (usually directional) between the distinctions in black:

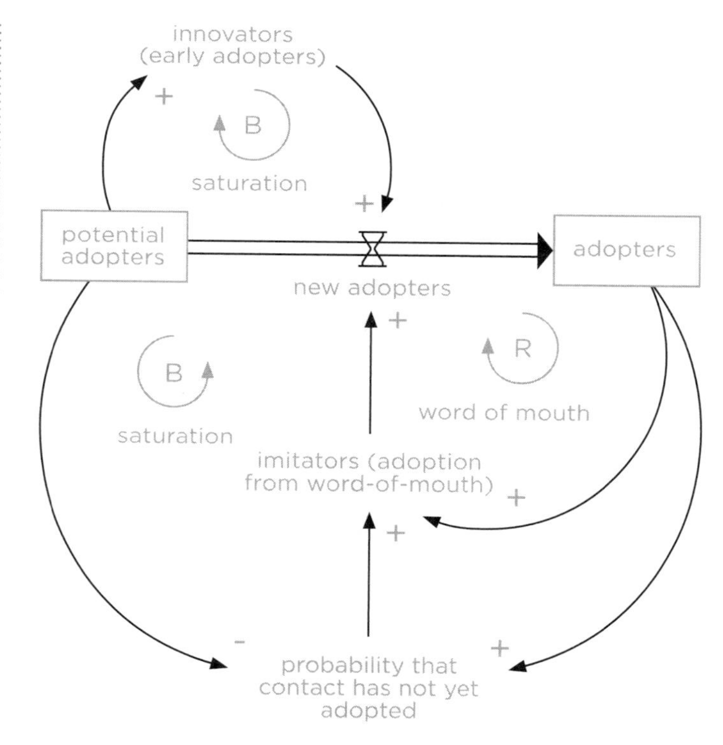

Figure 8.22: Relationships in System Dynamics Diagrams

The astute observer will notice that these directional relationships usually have also been made into distinctions (i.e., they've been "identified"). These identifications almost always take the form of "+" or "-" (increasing or decreasing). So in system dynamics relationships are specified (shown in Figure 8.23).

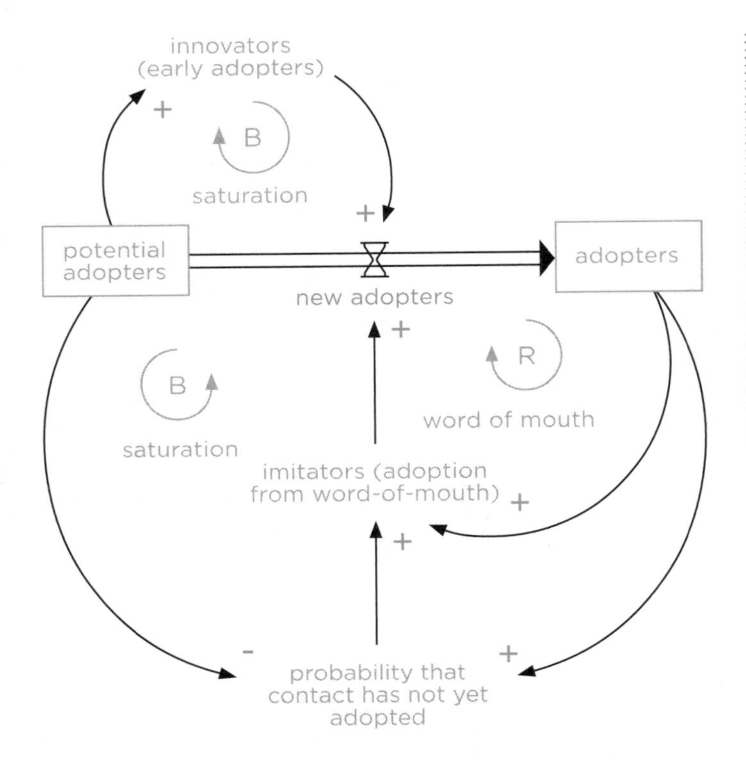

Figure 8.23: Identified (specified) Relationships

Now let's see where feedback comes in. Those relationships make up a lot of what system dynamics diagrams do. In fact, many "soft"[3] system dynamics diagrams do not include stocks and flows, just feedback loops. In the case of these soft system dynamics diagrams, all we are doing is chaining together a specific type of relationship (feedback) with various distinctions.

Feedback, which is so central to system dynamics diagramming, is really a part-whole system made up of specified parts and rela-

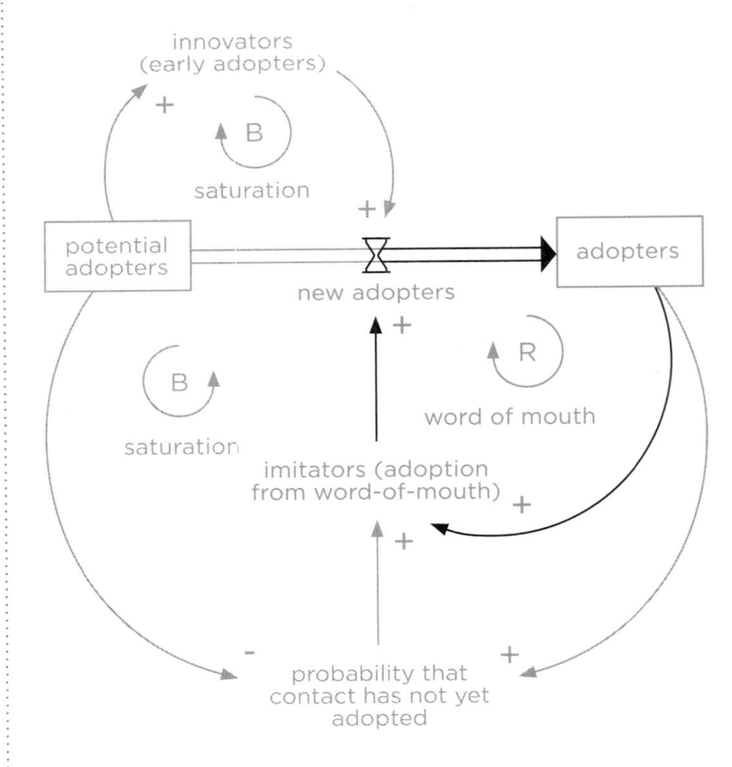

Figure 8.24: Feedback is a Part-Whole System of Relationships

[3]System dynamicists will often draw soft diagrams that include only feedback loops without stocks and flows or alternatively, diagrams that have no mathematical values behind them. These are sometimes called soft system dynamics diagrams.

tionships. You can see in Figure 8.24 that the "word of mouth" feedback loop is really nothing more than a system made up of three relational parts (black) between the structural parts (orange). The reinforcing feedback loop called "word of mouth" (green) is just a whole system made up of these parts (black and orange). In the diagram in Figure 8.24, there are two other such feedback loops that have the same part-whole structure. All feedback loops are systems of relational and structural parts.

Applying DSRP rules to system dynamics also highlights an important distinction between balancing and reinforcing feedback loops (Figure 8.25).

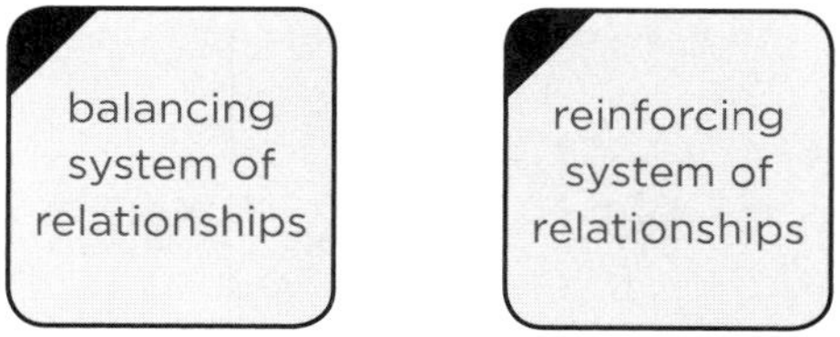

balancing system of relationships

reinforcing system of relationships

Figure 8.25: Distinction Between 2 Types of Feedback Loops

Stocks and flows are a core part of many system dynamics diagrams. Flows are just another distinct type of relationship. In quantitative diagrams, this relationship can take a number (a rate). This rate means the relationship is being distinguished as a particular number (no different than distinguishing it with a name). The stocks are accumulations (i.e., a part-whole system). The scale used in every stock is simply a fractional scale based on whatever the stock is (e.g., 30/100 potential adopters). Like all mathematical fractions, stocks are another example of a part-whole system.

So the underlying structure of all system dynamics diagrams is:

- Distinctions (labels, stocks, flows, feedback);
- Relationships (individual directional relationships, flows);
- Systems (stocks);
- Systems of relationships and distinctions: (systems of directional relationships, i.e., feedback loops); and
- Distinguished systems: (balancing vs. reinforcing loops).

Noticeably absent as a structural component of all system dynamics diagrams are perspectives. Of course, there is always an implicit or root perspective that belongs to the creator of the diagram. Also absent, and particularly important, is an explicit recognition of the other in the identity-other distinctions that make up all of the labeling in a system dynamics diagram. When the user is forced to test the validity of the constructs through distinction methods such as NONG/MECE (see chapter 3), systems thinking using DSRP is more effective. In addition, system dynamics diagrams could do a better job of showing the hierarchical structures that naturally occur in complex systems. Most system dynamics diagrams are flat, meaning all of the parts are laid out. There is no zooming. Yet hierarchies and the boundary conditions that enclose them are real-world structures that account for many important dynamics.

Rather than limiting ourselves to the specialized types of systems, relationships, and distinctions of system dynamics, we can include any number of other relationships, distinctions, and systems that might be relevant. Remember also that when we add perspective, everything changes.

SOFT SYSTEMS METHODOLOGY

Another popular systems approach called "soft systems methodology" (SSM) is straightforward, following seven steps. The seven-step process is somewhat linear, although steps 2-7 are repeating, as shown in Figure 8.26. The second step is based on a process given by the mnemonic device CATWOE, also shown in Figure 8.26.

Whereas system dynamics rarely takes perspective into account, SSM focuses on it. Specifically, SSM focuses on the process through which a facilitator guides stakeholders to build mental models (what SSM'ers call conceptual models) of real-world problems. The identified problem, the conceptual models of real-world systems, and the action taken to solve the problem are, in SSM, a participatory approach that includes stakeholders. It is therefore built on perspective-taking. However, the perspectives taken are almost exclusively

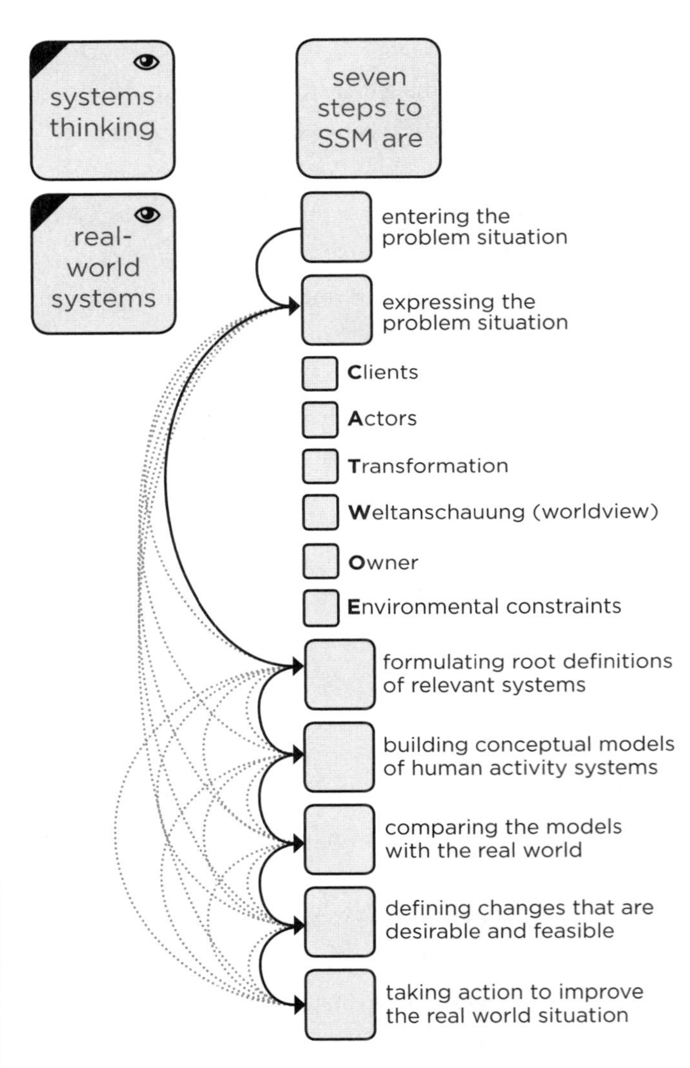

Figure 8.26: A Metamap of Soft Systems Methodology (SSM)

those of the participants/stakeholders. Unlike DSRP, there is no formal SSM axiom that promotes non-stakeholder (i.e., non-human) perspective-taking such as taking the perspective of a concept like evolution, costs, or manufacturing processes, etc. This small but important difference makes DSRP-style perspective-taking far more adaptive and robust.

One important distinction in SSM is that the seven steps are viewed from two vantages—systems thinking (individual's mental models) and the real world (reality).

If SSM was a sandwich, the bread would be Figure 8.26 with its seven steps. The meat of the SSM approach would be the process of creating mental models, which occurs throughout steps 1-7, but especially in steps 3 and 4. Here we see room for significant improvements to SSM. Figure 8.27 provides a typical example of SSM mental models, which are simplistic, qualitative networks of nodes and edges (usually with directionality). By developing less rudimentary conceptual models, the entire SSM sandwich would have more efficacy, efficiency, and effectiveness, terms SSM'ers use to assess the value of their process. The SSM group process can be a valuable tool to manage systems thinkers. Where the meat of the sandwich is concerned, SSM'ers would be better served by metamaps.

At its core, SSM is really a list of steps to take with a problem-solving group, and there are many equally useful models for doing this. For the novice facilitator, it helps to have steps and SSM's seven steps are a logical choice. But what you do inside those steps—the mental models the group builds—is most important. In addition, being able to think systemically (i.e., DSRP) allows the facilitator to adapt to the situation and develop whatever process is needed *in situ* for the complex system.

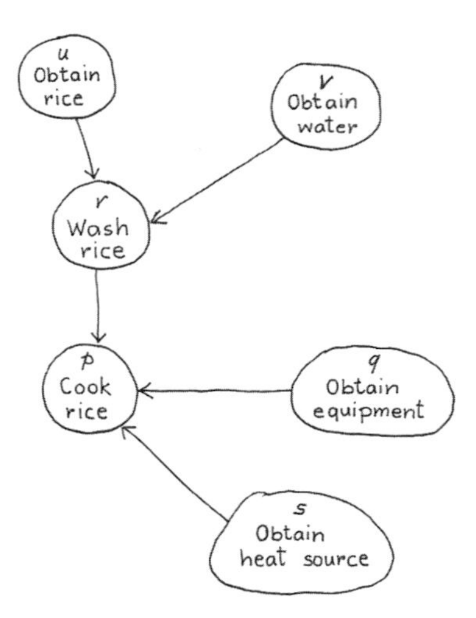

Figure 8.27: A Typical SSM Mental Model

NETWORK THEORY

Network theory is one of the most popular and powerful systems thinking methods. It is by far the most used, even if many of those who use it don't think of it as systems thinking. Network thinking is a useful tool for thinking about systems of all kinds:

physical, biological, chemical, ecological, neurological, psychological, social, economic, and so on. Let's explore the basics of network theory in order to show how a few simple additions can dramatically and profoundly impact network theory as a whole while also improving systems thinking.

Remember that visualization is critically important for human understanding. Network theory has provided an interdisciplinary tool for understanding complex systems in a visual way. Networks consist of nodes and edges (See Figure 8.28).

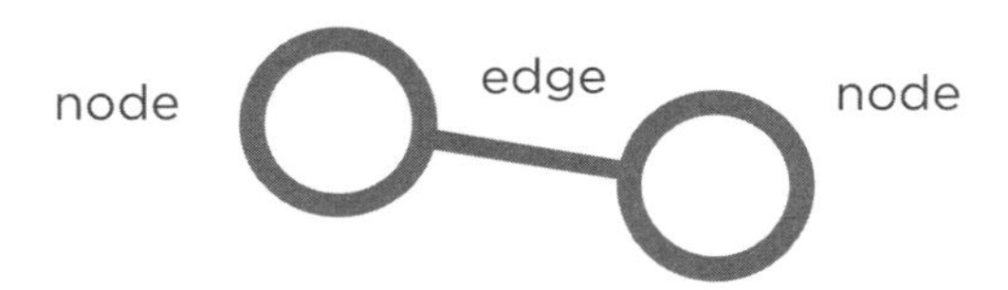

Figure 8.28: Basic Structure of Networks

The world is teeming with complex interactions. From this simple everyday inter-activity emerge complex webs of relationships. Some of the most fascinating and challenging systems are simply the result of many interactions. In these systems, the ways in which people or things are connected can dramatically affect the properties of the emergent system or network. Yet despite all this complexity, the simple node-edge construction of networks provides us with one of the most powerful visual-mathematical abstractions we have, capable of capturing the complexities of many real-world phenomena. Some connections are very strong like the relationship between predator and prey, other less so. But in most systems the pathways of interaction are crucial to the system behavior. The specifics of who can interact with whom affect the properties of the system and its internal processes.

Network theory explains complex systems of interaction using the simple premise that their parts are often connected to each other. For example, a network of websites and hyperlinks forms the world wide web, a vast and complex information network that is the backbone of the information age. Predator-prey interactions form an ecological network called a food web, the circuit diagram of an ecosystem. Complicated networks of human interaction form dynamic social webs, which are woven into the fabric of society. Financial markets, disease, weather systems, biological and ecosystems, concepts, languages, and economies are all networks. Even humans are nothing more than systems of networks: circulatory, renal, respiratory, neural, conceptual, etc.

Complexity theory holds that simple rules and local interaction underly complexity. Networks follow this theory because the simple micro-level interactions between nodes "add up" to become the large-scale behavior of the system. We want to be able to predict how the system will behave based on the simple underlying interactions, such as who or what connects to whom, and how many connections exist between and among the nodes. We see similar behavior across wildly different types of network structures, despite different specifics. This means that much of the behavior we see is occurring because of the structural and dynamic properties of the network, whether the nodes involved are as diverse as events, people, ants, cars, genes, cells, or computers.

EXTENDING NETWORK THEORY WITH DSRP

DSRP rules extend network theory in simple but profound ways. Some of these ways simply explicate ideas that are implicit in network theory already, whereas others add new structures and insights. In the text and figures that follow, we will walk you through the DSRP applications to network theory.

Start with a basic network of nodes and edges (See Figure 8.29). It could be any type of network, simple or complex. Think of this network as an abstract visual-mathematical representation for understanding systems.

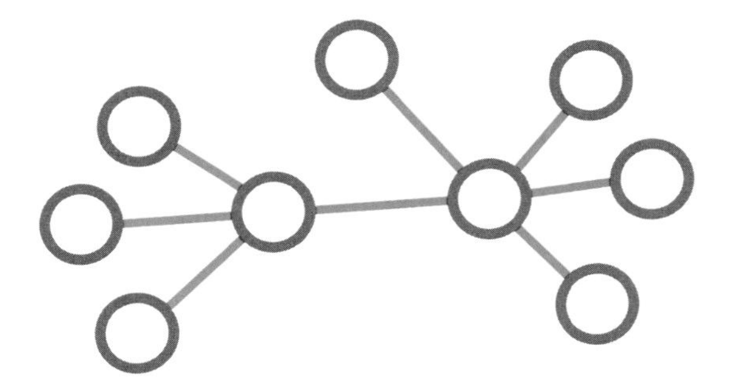

Figure 8.29: Nodes (Identities) and Edges (Relationships)

In Figure 8.30 we begin to extend basic network theory. Any node can be related to other nodes via edges (with or without directionality). But, by combining the distinction and relationship rules of DSRP, each of these edges could be its own node. Think for example of two computers linked together by a cord. Both computers are distinctly different things, but so is the cord itself.

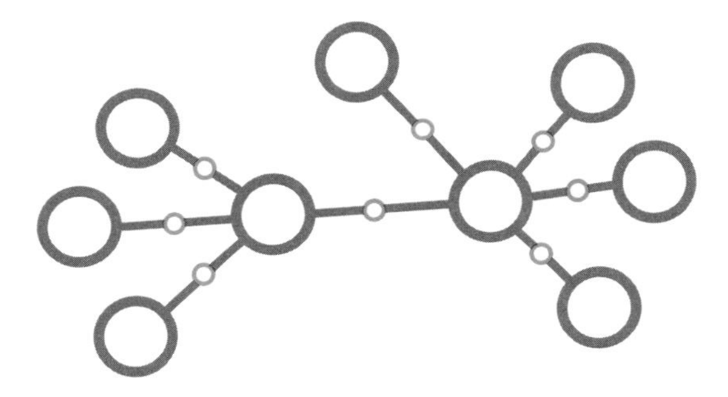

Figure 8.30: Every Edge Can Become a Distinct Node

Figure 8.31 shows another important extension of network theory that is either implicit or nonexistent. Any node (including edge-nodes we created in Figure 8.30) can be a whole system made up of parts with equal, lesser, or greater complexity relative to the whole. This is implicit in network theory for nodes themselves, and explicit in terms of clustering. It is important to note that this can occur to any node, including edge-nodes. What this means is that every node (large or small) in Figure 8.31 has the potential to be a complex system made up of many related parts, and so on.

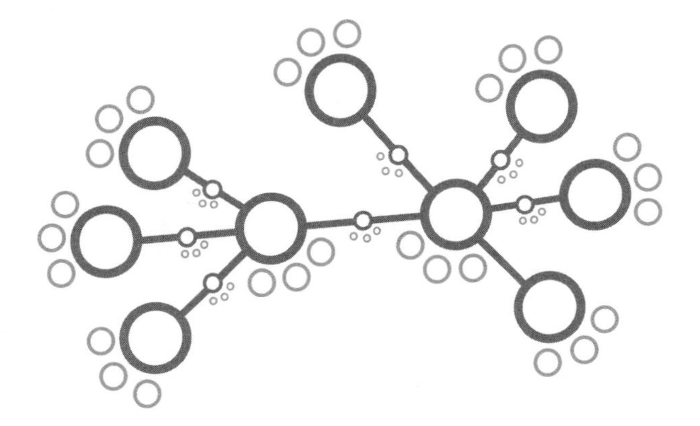

Figure 8.31: Every Node and Edge Is a Part-Whole System

Any node (or system of nodes) can be the point or the view (eyeball) of a perspective. This means that every node in Figure 8.32 (which is everything) can be the point of a perspective, making all the other nodes the view. This is not explicit in network theory.

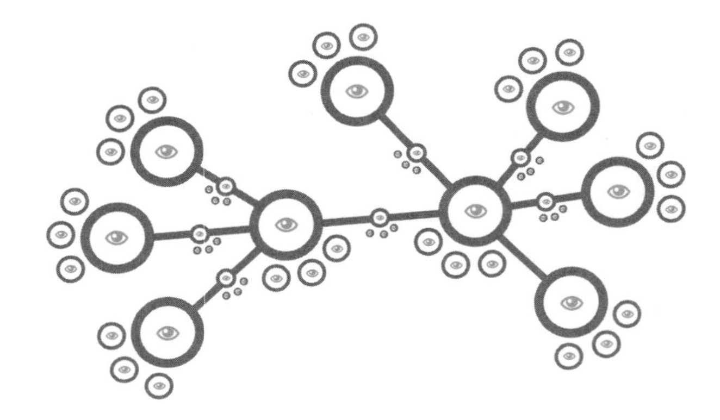

Figure 8.32: Any Node or Edge-Node Is a Point or a View in a Perspective

Every node is distinguished by the other nodes it is with. This is not explicit in existing network theory. Figure 8.33 shows that any node (an identity) is defined not only by what it is, but in contrast to system of things it is not (in red). This is a new insight (scale-free boundary conditions) that is often overlooked in network theory and essential to systems thinking.

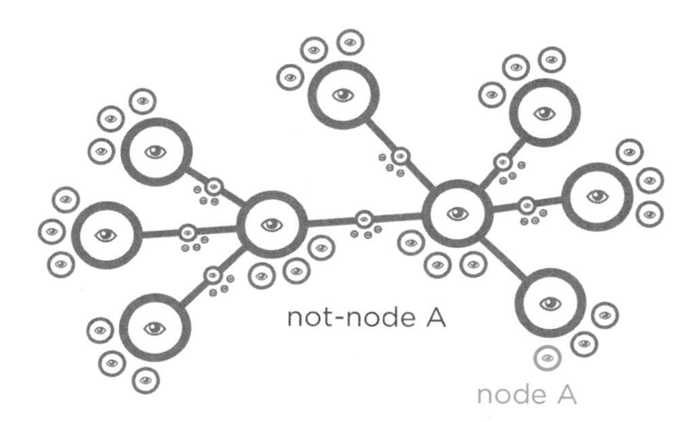

Figure 8.33 Every Node Is Defined by the Other Nodes

If the DSRP rules were fully explicated on all nodes and edges of a network (as shown in Figures 8.29-8.33) the maximum potential complexity of the network would be reached. In other words, no matter how complex the system gets (in theory at least), we would have a conceptual model capable of capturing it.

Of course, in the same way that all nodes in a network do not need to be related (have an edge between them) even if they could be, not all the nodes in a system need to manifest every DSRP possibility. They are just representations of what is possible. In practice, all nodes, all edges, and all rules would require the power of a supercomputer and countless variables would remain unknown. Nonetheless, since DSRP rules elucidate what should exist structurally they have predictive power.

What would a network look like that has the extended properties we have outlined? Facebook. Because Facebook is a social network, it's easy to understand that it is perspectival. Each person's view of Facebook is unique. For example, when we all visit www.apple.com, we all see the same webpage. But when we point our browser to www.facebook.com, we all see a webpage that is from the unique perspective of our profile. So although our group of Facebook friends represents a traditional node-edge network, we can also see that every one of the nodes has a unique perspective on the network. None of the nodes (people) in the network are seeing a bird's eye view of the Facebook network, we are each seeing a tiny slice that includes the people we are connected to. We also can see that each profile is a rather large whole system made up of many parts that are stored in a large data table (all our friends, family, groups, pages, images, posts, shares, likes, etc.). There's also a less known feature in Facebook called relationships. You can go to any friend's profile page and click on the top right to "see relationship." This allows you to see a uniquely generated page that represents the relationship between you and your friend. It combines all the parts that you share into a distinct page. This relationship page has a unique identity and a unique URL that combines the names of you and your friend. This means that the relationship is a distinction but it is also a system.

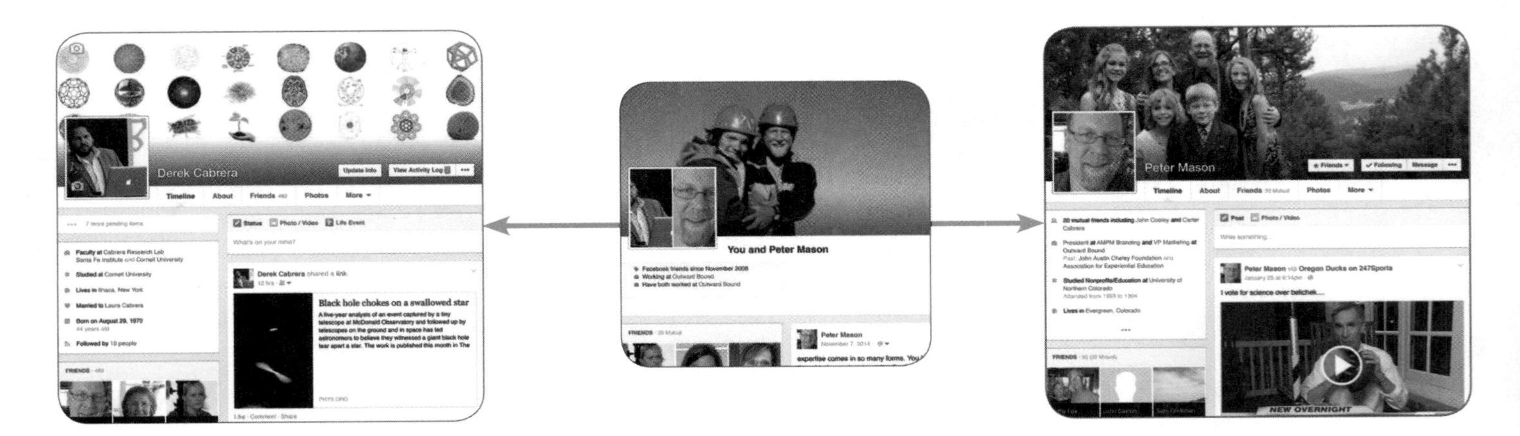

Figure 8.35: Edge as Node (Identity) and System

You can see this in Figure 8.35, which shows the relationship between Derek and one of his old climbing buddies, Peter. On the left is Derek's profile page; on the right is Peter's. In the middle is the relationship between Derek and Peter (from the perspective of Facebook!), which includes all of the parts of their relationship: comments, things they've both shared, photographs from past climbing expeditions, and mutual friends and workplaces. What we can see is that Facebook is a network that is based on distinctions (every identity is different and contrasted with others), systems (every profile is a system made up of many parts), relationships (shared items and interactions between profiles, likes, comments, etc.), and perspectives (every profile is a unique perspective on the system).

It is easy to see this on Facebook, because it's a social network where all the people have eyes and brains. But can we see that the same dynamics are at play in an ecological web? Do we grasp that the same underlying structure and dynamics are true for a computer network, a gene network, an author citation network? Or that each organism, computer, gene, or cell experiences the ecosystem from their unique perspective and makes decisions, builds relationships, and behaves based on this perspective? Can we see that within each interaction between any two things is an entire world of complexity? Do we understand that even for inanimate things like ideas, atoms, and words, our mind utilizes perspective to understand?

When we understand that everything can be imagined in terms of networks, that's a big step. But when we see that any system can be thought of in terms of networks following DSRP rules that's quite profound. Nothing can get more complex than a network of individual nodes that are co-defining each other, relating and not relating to each other, assembling into and out of part-whole groupings, and taking perspectives on slices of the whole. When we take into account that, like Facebook, every relationship has the potential to also be a distinction and a system capable of containing immense complexity, our thinking aligns with the way the world actually works. Complex? Yes. But simple underneath.

UNIVERSALITY AND PLURALISM

The vast plurality of informal and formal systems thinking approaches, methods, ideas, and applications make sense in light of the universality of DSRP. That one simple set of rules underlies such a vast number of powerful approaches means that for the motivated systems thinker, more can be understood and accomplished with less. For the champion of systems thinking—who believes that it offers humanity a new hope for solving wicked problems and tackling everyday tasks—universality provides a foundation upon which to build.

SECTION 3

7 BILLION SYSTEMS THINKERS

SECTION 3
7 BILLION SYSTEMS THINKERS

Systems thinking can be applied to many things across the spectrum of human knowledge and experience, from atoms to molecules and from ecologies to economies. Yet systems thinking is at its core a human endeavor. Systems thinking is done by individuals who want to solve both everyday and wicked problems. When individuals alone cannot solve their problems, they form organizations. Organizations need not be formal IRS entities like corporations or nonprofits. They can also be loose networks, affiliations and partnerships, families, and friendships. Organizations can span from the smallest (2 people) to the largest (millions of people). But people in and of themselves do not constitute an organization. That is, two people or ten people working in the same building do not make an organization. There is another ingredient that's required for this recipe: a shared mental model.

In the next two chapters we will share our experience with developing individuals into systems thinkers, and organizations into systems thinking organizations. As an individual, you'll see that there are some pretty valuable emergent properties that result from doing systems thinking. And if you have an organization or want or need one, we provide a recipe for ensuring that it will be a systems thinking organization. First, start with systems thinkers or teachable individuals. Next, develop a shared mental model of a very special kind of CAS-based Vision and Mission. Then build your entire culture around it. Developing individuals into systems thinkers is part and parcel of developing systems thinking organizations.

CHAPTER 9

SCALING SYSTEMS THINKING

INDIVIDUALS AND ORGANIZATIONS

Chapter 1 explained that systems thinking gives new hope for solving both everyday and wicked problems. Chapter 2 revealed the purpose of systems thinking: to increase the likelihood that our mental models accurately match how real-world systems work. This feedback loop between our mental models (how we think things work) and the real world (how they actually work) is the cornerstone of systems thinking, science, evolutionary processes, adaptation, and learning (). In turn, individual and collective learning drives the evolution of culture, organizations, and society. A complex adaptive system (CAS) results from independent agents operating on simple rules that, based on the collective dynamics among the agents, cause the global behavior of the system to emerge. The adaptive part is that CAS learn from and adapt to the environment around them. Chapter 3 explained that systems thinking is an emergent property of four simple rules called DSRP that act upon information. In chapters 4 through 8, we explored DSRP through practical tools for modeling systems thinking and case-based examples.

It is safe to assume that your desired use of systems thinking will follow one or two applications:

1. To develop systems thinking skills in yourself or another person (a child, student, employee, etc.); or
2. To lead a systems thinking organization of some kind (whether it be a family group, a for-profit business, a nonprofit, or governmental initiative).

In this next section, we apply these new insights to the creation of systems thinkers and systems leadership for organizations. If you're interested in developing systems thinkers, whether yourself or another, then develop your and their practice of DSRP. The best way to develop systems thinkers is to practice the DSRP rules until they become second nature. If you want to pursue a role as a systems leader and apply systems thinking in your organization, you must first understand the role systems thinkers play as agents in the organization. Second, you must learn four additional simple rules: vision, mission, culture, and learning, or VMCL. Figure 9.1 illustrates the emergent property that results at both the individual and the organizational level. In other words, individuals following DSRP rules become systems thinkers, and these systems thinkers following a new set of rules (VMCL) produce optimally successful and adaptive organizations. The first step to become a systems thinker is to apply the simple DSRP rules of systems thinking to oneself. We all work with others in groups in some way. Therefore, after becoming a systems thinker the next step toward having a greater impact in the world is through an organized group of systems thinkers.

We all seek optimal individual and organizational outcomes—levels 3 and 6 in Figure 9.1. Ironically, our best chance to achieve these macro-level or emergent outcomes is to focus our efforts on individual agents and simple rules—levels 1 and 4 in Figure 9.1. In this sense we build, albeit indirectly, better humans and organizations from the bottom up.

The principle takeaway from this book should be that if you want to apply systems thinking to individual or organizational problems you will need to focus on the simple rules to bring about the change you seek. So, for example, Chapter 10 explores the idea that if the outcome we want is systems thinkers, we should focus our energies and efforts on applying the simple DSRP rules to individuals. Likewise, Chapter 11 shows that systems thinkers, banded together as an organization, must follow another set of simple rules called VMCL, which leads to a truly adaptive and robust organization.

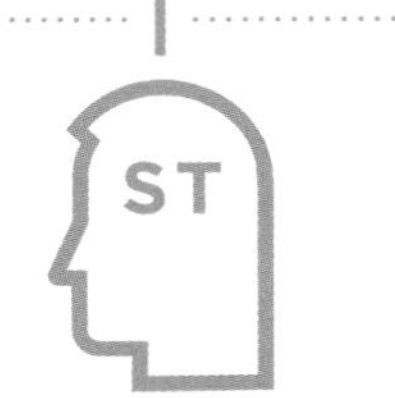

FIGURE 9.1: Bottom-Up Influencing Individual and Organizational CAS

To reiterate, Figure 9.1 illustrates where to focus our efforts if we are interested in: (1) becoming or developing a systems thinker, or (2) developing a systems thinking organization. We can think of these two goals using a popular concept called circle of influence versus circle of concern, illustrated in Figure 9.2. The most effective people are those who are proactive rather than reactive. Proactive people focus where they have influence (which for most of us is a smaller circle than our circle of concern). When we focus on our influence, our influence grows. In contrast, a reactive approach focuses on concerns, which in turn detracts from our influence. Because emergent properties are robust, they are often what we want, but because they are born of collective dynamics and self-organization, we rarely have control over them. However, we can influence the simple rules and the agents and alter the complex emergent properties (system behaviors and outcomes) that so often fall in our circle of concern.

SCALING SYSTEMS THINKING

Business leaders talk about scaling a lot. It means taking a single or local success and making it accessible to a larger audience. Of course, this could be done for various reasons, to increase profit or to increase the collective good, or both. Scaling used to be called democratizing. Our founding fathers democratized personal agency, liberty, and freedom. Henry Ford helped to democratize the automobile by producing it at a price point that was accessible to average

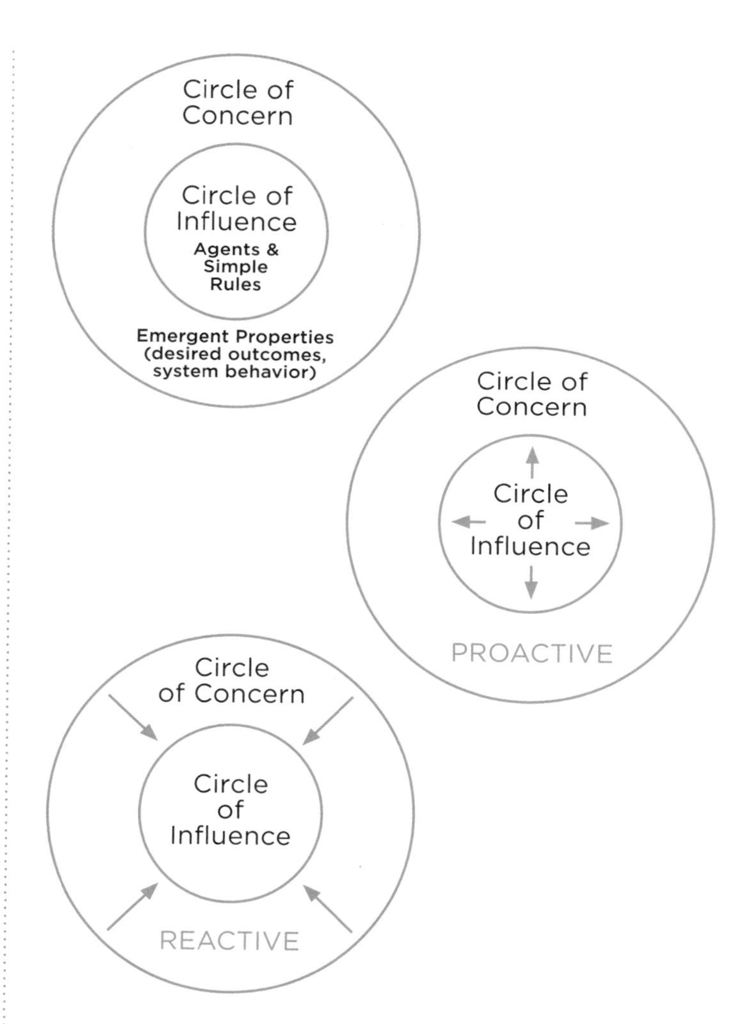

FIGURE 9.2: Influence Vs. Concern

people. Bill Gates and Steve Jobs democratized the personal computer. Marc Andreessen[1] democratized the internet. John Sylvain democratized those little single-serve coffee pods that now crowd landfills with enough non-biodegradable waste each year to wrap around the globe 10.5 times.[2] Not everything needs to be scaled, but systems thinking does.

Our goal in writing this book—and in all the work we do—is to democratize systems thinking. Democratizing (or scaling) systems thinking is necessary because systems thinking as a concept is a lot like democracy. You can't do democracy in an aristocratic way. Democracy has to be of, for, and by the people. Not for the people but of and by the ruling class. We are seeing the dangers of this kind of aristocratic democracy today in government bureaucracies where ostensibly democratic processes are highly manipulable by the few for their own ends (e.g., Wall Street banks and K Street lobbyists). For systems thinking to be truly revolutionary it needs to be of, for, and by the people.

There's a new kind of interpretation of how systems thinking is useful that not only misses the point of systems thinking but can also be quite dangerous. It's the idea that the purpose of systems thinking is to develop a comprehensive, big-picture view of large-scale systems so that leaders can command and control these systems better. This view evokes a military general perched atop a hill surveying the battle below. This is dangerous because it focuses our energy in the place where we will be least effective and also misses the point concerning how we can better design more robust systems.

We've seen organizational leaders and policymakers fall into this pitfall and it's often tragic, because their intention was to get out of a hole and they end up digging a deeper one. Let us explain through a powerful case that is quite literally a life and death example of what it looks like when we make the mistake of thinking about organizations from a top-down perspective rather than as a CAS. Figure 9.3 shows a multi-million dollar map developed by the CIA and Pentagon for the war in Afghanistan. It was so expensive and controversial it was featured in the *New York Times*.[3] It's a System Dynamics map showing many feedback loops.

1 Metcalfe, B. (1995, August 21). Microsoft and Netscape open some new fronts in escalating Web Wars. *InfoWorld*, 35.

2 Kane, C. (2015, March 17). 10 inventors who apologized for their inventions. Retrieved April 16, 2015, from http://fortune.com/2015/03/17/10-inventors-who-apologized-for-their-inventions/

3 Bumiller, E. (2010, April 26). We Have Met the Enemy and He Is PowerPoint. *New York Times*. Retrieved April 17, 2015, from http://www.nytimes.com/2010/04/27/world/27powerpoint.html?_r=0

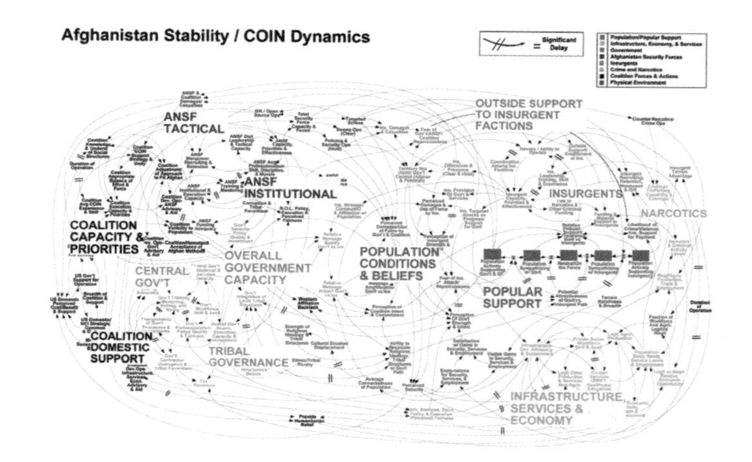

FIGURE 9.3: God's Eye View

Here's what General Stanley A. McChrystal, the guy who was supposed to be implementing the strategy, said about the map: "No one can actually figure the darn thing out. When we understand that slide, we'll have won the war."[4] Brigadier General H. R. McMaster warned, "It's dangerous because it can create the illusion of understanding and the illusion of control."[5]

In contrast, informed by a CAS perspective, Four Star General George W. Casey, Jr., who as a systems thinker coined the term "VUCA world" (an acronym for one that's volatile, uncertain, complex, and ambiguous) explains,

> *It's essential to distill a message into a few key points and hone its delivery. Clear communication is like sharpening a pencil: You slowly remove the unnecessary until you are left with a pointed, useful message. After trying and failing to communicate our strategy, we went back to the drawing board and came up with "Clear. Hold. Build." and re-communicated it. It was much better understood.*[6]

Clear. Hold. Build. Now that's a clear mission.

FIGURE 9.4: Agents and Simple Rules View

Figure 9.4 shows that General Casey's 3 simple rules can be applied at multiple levels of scale across his organization. A platoon of soldiers tasked with clearing a house can apply the mission to clear the house of dangers, then set up security measures to hold the house, and then build relationships in the house to ensure longer-term success. At every level of scale upward, from the neighboring block that the house is

[4] Ibid
[5] Ibid

[6] Hogge, T. (2014, November 19). What I Learned About Leadership From A 4-Star General. *Business Insider* Retrieved May 16, 2015.

on to the neighborhood, town, region, and eventually state can utilize the same 3 simple rules.

What Casey understood is that systems thinking isn't just something that generals do to get the big picture, it's something that all the agents (e.g., soldiers and commanders) do at every level in the organization. He needed to help all the agents identify the simple rules that collectively led to the emergent properties they were all seeking.

In the chapters that follow, we will explore how the simple rules of DSRP lead to far more complex and robust outcomes than we might expect: not only the ability to problem solve but also creativity, emotional intelligence, and a deeply humanistic orientation. In short, following the DSRP rules causes adaptive, robust, lifelong learning to emerge. In chapter 11 we'll explore a systems leadership model for organizations that relies on the collective dynamics of these learners to create a learning organization that is laser focused on the outcomes the organization wants.

CHAPTER 10

SYSTEMS THINKERS WANTED

SYSTEMS THINKERS WANTED

Nietzsche[1] developed a dichotomy between Apollonian and Dionysian types of thinkers, which Gell-Mann referred to at the founding of the Santa Fe Institute (SFI). He wrote:

> There are some psychologists and pop psychologists who like to place people on a scale running from Apollonian to Dionysian, where, roughly speaking, Apollonians tend to favor logic, rationality, and analysis, while Dionysians go in more for intuition, feeling, and synthesis. In the middle are those tortured souls, the Odysseans, who strive for the union of both styles. The new institute would have to recruit a number of Odysseans to be successful![2]

Although the typology is far from scientific, we often make a distinction between these two types of thinkers, perhaps for the sake of discussion or perhaps because we all feel a tugging or yearning for both.

[1] Nietzsche, F. (2000). *The birth of tragedy*. Oxford: Oxford University Press.

[2] Gell-Mann M. The Concept of the Institute. In: Pines D, ed. *Emerging syntheses in science: proceedings of the founding workshops of the Santa Fe Institute*, Santa Fe, New Mexico. Redwood City, Calif.: Addison-Wesley; 1988. pp 8.

Early (and even some contemporary) systems thinkers fell victim to the dichotomy by pitting holists against reductionists, as if one can be considered without the other. Or there is the left-brained (logical) and right-brained (creative) dichotomy, which is accepted as fact in the public sphere despite numerous scientific debunkings.[3]

Another popular dichotomy is hard versus soft skills where logic and analytics is thought to be hard-edged and creativity and intuition are touchy-feely soft skills. Yet this hard-soft metaphor is often misinterpreted to mean that hard skills are more difficult and therefore more valuable and soft skills are easier to develop and therefore devalued as easy, which is untrue.

Others dichotomize with the metaphor the heart and mind. The Dalai Lama writes "It is vital that when educating our children's brains that we do not neglect to educate their hearts."[4] Yet the original Dalai Lama and founder of Buddhism, the Hindu Prince Gautama Siddharta, said, "We are what we think. All that we are arises with our thoughts. With our thoughts, we make the world."[5] Many of the things we think of as occurring in the heart are actually occurring in the brain, so the heart-brain dichotomy gives us this false sense that the brain is a cold, calculating place

[3] Nielsen, J. A., Zielinski, B. A., Ferguson, M. A., Lainhart, J. E., & Anderson, J. S. (2013). An Evaluation of the Left-Brain vs. Right-Brain Hypothesis with Resting State Functional Connectivity Magnetic Resonance Imaging. *PLoS ONE*, 8(8), e71275. doi:10.1371/journal. pone.0071275

[4] Dalai Lama (twitter communication, December 5, 2011).

[5] Byrom, T. (1976). *The Dhammapada: The sayings of the Buddha* (p.XV and 1). New York: Knopf.

capable only of analytical rationality and the heart is our warm and fuzzy emotion center. What we think of as the emotional, motivational, compassionate, or heart-full side of humanity—things like ethics, morals, values, etc.—are integrated with cognition into a thing called the "mind." Our mind processes our internal world—our motivations, thoughts, and feelings—and it is through a reflection and awareness of these elements of the mind that we come to better understand ourselves and others.

The duality yin and yang are more integrated than dichotomous as yin and yang are separate but included in and leading to each other (☯). The critical thinking, executive function, analytical, intellectual, and cognitive sides of the human are *yang* and the emotional, prosocial, self-aware, intrapersonal,[6] and interpersonal sides of being a human are *yin*. The Tao refers to the middle way between them.

The truth is that no matter which terms we use to describe the two sides, the dichotomy is false. We want to honor both (or perhaps not see them as distinctly different in the first place), and possibly even indicate their mutual dependence and complementarity. While these two thought styles may appear to be qualitatively different, they are also highly complementary. Perhaps Gell-Mann's suggestion is the right path, to think of those who possess both thought styles as Odyssean thinkers.

[6] Intra-personal intelligence refers to knowledge, skill or ability in managing one's relationship with oneself, whereas interpersonal intelligence refers to managing relationships with others.

SYSTEMS THINKERS WANTED

The Odyssean thinker is a systems thinker. It is not only the kind of thinking we want for ourselves in terms of our own personal and professional development, but also for our children. Whether as parents or educators, we want our children and our students to grow up to be Odyssean thinkers; well-rounded, thoughtful, whole human beings make good citizens. It's also what business leaders want in their prospective employees.

It is important that we don't see systems thinking as merely developing the analytical self (Apollonian, left-brain, yang thought style), but also as developing the social-emotional self (Dionysian, right-brain, heart, yin thought style). Systems thinking is a tool to better understand complex systems of any kind, which include our internal selves, emotions, behaviors, and motivations. It's important to develop mastery of more traditional "intellectual" content, but without a parallel development of the self, we will end up with a room full of smart people who cannot relate to one another, cannot work in teams, and who make decisions based on ego rather than discovery and contribution. In short, we will end up with functional psychopaths. This result is obviously unacceptable (although increasingly prevalent[7]) and is in

[7] Bate, C., Boduszek, D., Dhingra, K., & Bale, C. (2014). Psychopathy, intelligence and emotional responding in a non-forensic sample: an experimental investigation. *The Journal of Forensic Psychiatry & Psychology*, 25(5), 600-612. doi: 10.1080/14789949.2014.943798.; Stevens, G., Deuling, J. K., Armenakis, A. A. (2012). Successful Psychopaths: Are They Unethical Decision-Makers and Why? *Journal of Business Ethics*, 105(2), 139-149. doi: 10.1007/s10551-011-0963-1.

direct contrast with what we all know we need and want in ourselves and in the humans around us.

It is important that we don't see systems thinking as merely developing the analytical self (apollonian, left-brain, yang thought style), but also as developing the social-emotional self (dionysian, right-brain, heart, yin thought style).

When we're giving talks around the country to educators, policymakers, parents, and employers, one of the first things we ask them is to make a list of the qualities they want children to possess as adults.

As parents, teachers, policymakers, and employers we are surprisingly all on the same page about the kinds of adults we want. No matter how many times we ask different groups in different parts of the world to make their list, it almost always comes out the same (see Figure 10.1). To summarize, we want adults to be:

- Critical thinkers who can analyze and solve problems;
- Creative thinkers who can see new and innovative solutions to problems;
- Scientific thinkers who can recognize biases, seek new knowledge, and investigate complex phenomena;
- Prosocial thinkers who can work well with others and build strong communities; and
- Emotionally intelligent individuals who possess a sense of themselves and what they offer to the world.

PROBLEM SOLVER **SMART** OPEN **ETHICAL**
KIND **PRODUCTIVE** GOOD COMMUNICATOR
SELF-AWARE WILLING LISTENERS **CREATIVE**
FLEXIBLE **COOPERATIVE** HARD WORKING
COMPASSIONATE CURIOUS **RESPONSIBLE**
SELF-RELIANT **PERSEVERING** THOUGHTFUL
COLLABORATIVE HAPPY **REFLECTIVE**
RESILIENT **CRITICAL THINKERS** SKEPTICAL
SELF-REFLECTIVE ADVOCATE **ANALYTICAL**
PERSISTENT **HONEST** SENSE OF HUMOR
ADAPTABLE SKILLFUL LEARNERS **INSPIRED**
INSPIRING **ACCEPTING** ACCEPTED **PASSIONATE**

Figure 10.1: List of Qualities of a Well-Rounded Person

But how can we accomplish this critical task? How can we develop ourselves to be Odyssean systems thinkers? How can we develop our kids to be the kinds of adults we seek for the future? How can we train or retrain employees to become systems thinkers to meet the demands of a complex world?

Systems thinkers are needed in every business, home, and classroom. Yet we don't seem to be creating more of them. This is because systems thinkers are not born or made, they emerge. Because a systems thinker is an emergent property, we simply need to teach the simple rules of DSRP. Over time the patterns that emerge will be that of a systems thinker: a complex, adaptive, Odyssean thinker and lifelong learner.

Throughout this book we've been talking about the mind. From problem solving to mental models to information and structure to cognitive jigs, etc. So you likely have a sense now that DSRP can significantly increase one's analytical, intellectual, and cognitive abilities. But it is perhaps less obvious how practices that are cognitive, analytical, and intellectual apply to the emotional, social, self-reflective side of our human existence. Developing a balanced thought style between the needs of the self and the other, between one's emotions and one's cognition, and between intrapersonal and interpersonal awareness is the key to being a well-rounded person. These are the kind of people we want our children and students to become and our friends, employees, colleagues, and leaders to be.

Can the simple DSRP rules of systems thinking help us to develop emotional intelligence, prosocial skills, compassion, empathy, introspection, perseverance and grit, and an internal ethical compass? The answer is yes and the key is an idea called *metacognition*.

DSRP INCREASES METACOGNITIVE AWARENESS

The scientific term for thinking is cognition. The scientific term for thinking about one's thinking is metacognition. Most people think of metacognition as an increase in awareness in general: awareness of one's thoughts, emotions, and motivations, but also awareness of how one interacts with the outside world. Much of the discussion in this book has been about metacognition. Systems thinking is a particular type of metacognition that focuses on and attempts to reconcile the mismatch between one's mental models and how the real world works. The following are all acts of metacognition:

- Awareness that everything you experience is a mental model that approximates (either poorly or well) the real world;
- Awareness of the distinctions you make and the emotions and motivations that may have influenced you to make them;
- Awareness that everything you and others think is being influenced by one or more perspectives;
- Awareness that there are many ways to organize and interrelate ideas and things (and your current way is just one of them); and
- The ability to distinguish among cognition (thinking), emotion (feelings), and conation (motivations) and the awareness of how these different human capacities influence our mental models and behavior.

These are all metacognitive acts. Research has shown that metacognition has many positive benefits. Take a few findings from recent research in the field of metacognition.

Increasing metacognition early in life builds a stronger brain with more gray matter, which is correlated with intelligence and cognitive capacity.[8] Encouraging metacognition, or increasing a student's own awareness of his/her learning processes, leads to an increase in integrative abilities and transfer across areas of study.[9]

These increases in cognitive capacity, brought on by metacognition, are also highly correlated with one's potential for emotional intelligence. The most comprehensive research on emotional intelligence—which companies are currently spending millions to develop in their employees—concluded that metacognitive ability was necessary for emotional intelligence because, "emotion perception must causally precede emotion understanding, which in turn precedes conscious emotion regulation and job performance."[10]

[8] Weil, R.S. and Rees, G. (2010). Decoding the neural correlates of consciousness. *Current Opinion in Neurology*, 23:649–655

[9] Huber, M. T. & Hutchings, P. (2004). *Integrative learning: Mapping the terrain*. Washington, D.C.: Association of American Colleges and Universities.

[10] Joseph, D.L. and Newman, D.A. (2010) Emotional intelligence: An integrative meta-analysis and cascading model. *Journal of Applied Psychology*, Vol 95(1), 54-78.

Can the simple DSRP rules of systems thinking help us to develop emotional intelligence, prosocial skills, compassion, empathy, introspection, perseverance and grit, and an internal ethical compass? The answer is yes and the key is an idea called metacognition.

Somewhat surprisingly, what this means is that metacognition has a more profound effect on emotional intelligence than does direct instruction in emotional intelligence.

In other words, teaching metacognitive skills will yield higher emotional intelligence.

Finally, research also shows that, "insights into our own thoughts, or metacognition, is key to high achievement in all domains."[11] Why does metacognition work? It is analogous to the idea that we should teach how to fish rather than provide fish. Metacognition empowers because it provides awareness of one's own thought processes. When we focus on metacognitive awareness of the simple rules of systems thinking, we equip humans with the ability to understand everything they encounter, not just science or math, but any and all systems in life, including one's own destructive pat-

[11] Fleming, S. M. (2014, September 1). Metacognition Is the Forgotten Secret to Success. *Scientific American Mind*, 5(25), p32

terns and the perspectives of oneself and others, all of which are important for conflict resolution.

The development of self-awareness (or metacognition) exists on a continuum shown in Figure 10.2. We begin at the unconscious incompetence stage, which means we don't know what we don't know. We can't comprehend what self-awareness would be like, or the value it would bring to our daily life. If we are lucky, something "wakes us up" and causes a search for something more. Next, we move into the conscious incompetence stage where we realize we have something we need to learn, and are motivated to learn it.

It takes some time to develop competence in the skill. Once some competency is acquired, we are in the unconscious competence stage where the skill is being practiced regularly, but the user may not fully be aware of the skill. There is cognition, but not metacognition. We're good at things, but we don't always know the processes that create our competencies. Moving to the conscious-competence stage means becoming aware of what you are doing so that you can adapt what you are doing whenever you need to.

If you've had any experience with public schools, private schools, or Montessori schools, you may have seen something like this continuum in action. We had the opportunity to work with many such schools while researching our book, *Thinking at Every Desk.*[12] It is a foreign idea in many schools that kids should think rather than memorize. Why not continue to cover the information as we have always done? In many districts, they believe it is innovative, instead of ironic, that their new initiatives include getting their school to focus on learning. So most public schools are in the unconscious incompetence stage. You need to actually convince some teachers, principals, and superintendents that thinking is important! Some private and public schools that have chosen to be alternatives to traditional schooling have embraced the importance of getting kids thinking, and are in the process of learning how to do it. Montessori schools embraced the importance of thinking skills long ago, so they've had some experience doing it. They are in the unconscious competence stage. They are very good

Figure 10.2: Consciousness and Competence Continuum

[12] Cabrera, D., & Cabrera, L. (2012). *Thinking at every desk: Four simple skills to transform your classroom.* New York, NY: W.W. Norton.

at getting kids to think, but they are unaware of how to get kids to develop awareness of their thinking. Moving to the conscious competence stage is akin to competitive athletes becoming aware of their breathing, gate, nutrition, and biomechanics in order to help them improve their game while in the flow of competition.

Awareness of one's thinking applies not only to curricular content taught in school, but also to awareness of oneself, and our relationship to others around us. That is the essence of character, ethics, and an internal compass. For example, just because you can, doesn't mean you should. Just because I can [insert any bad behavior here] and get away with it, doesn't mean I should. You could insert any of the following and many more: lie, cheat, steal, hurt, get more than my share, manipulate. Sometimes a simple idea can pack a lot of punch. Let's take a look at the 4 simple rules of DSRP to see the underlying ethical stance of each rule. Implicit in this exercise is the belief that practicing these DSRP rules will produce people of character.

Remember that each DSRP rule co-implies two elements (e.g., systems are composed of parts and a whole). So you will see that in each case (D, S, R, and P) utilizing these rules not only leads to powerful Apollonian (logical) but also Dionysian (intuitive) people. You will see over and over again that the four patterns of DSRP and their co-implying elemental structure actually provide for metacognition and will result in an Odyssean thought style—an and/both rather than either/or mindset.

THE ETHICAL STANCE UNDERLYING MAKING DISTINCTIONS

Every distinction, whether it is something innocuous like "cup" or meaningful like "love," creates a boundary between what meaning lies inside and outside the idea. Sometimes these boundaries, like the things they define, are innocuous and do no harm. But other times these boundaries can be oppressive to the other.

We make countless distinctions every day in every domain of our lives. In math, for example, we distinguish between an expression and an equation. In science, we distinguish between solids, liquids, and gases. In adolescent development, we distinguish between risk and positive behaviors. Introspectively, we distinguish between thoughts and feelings. Distinguishing between one's thoughts and feelings is core to emotional intelligence and regulation (an executive function). We need to know the difference between the two. We need to experience our feelings but not project them onto others, a fine-grained distinction itself. We need to be reflective about our feelings so we learn to distinguish stimulus from emotional response. We need to gauge and distinguish between situationally appropriate and inappropriate emotional responses. Out of this basic distinction-making process many desired outcomes emerge: healthy coping mechanisms, resiliency, ego-checking, self-awareness, and self-regulation.

SEEING THE OTHER

The majority of our language focuses on the thing/idea part of distinctions but fails to explicitly recognize the other. When we describe a child who has Attention Deficit Disorder (ADD), for example, we are defining that child in relation to the norm. Again, we could just as easily diagnose the normal child with BTD, or "boredom tolerance disorder." You can imagine that if you were suddenly labeled boredom-tolerant you might start to think of that label as being quite negative, insinuating a lack of intelligence and creativity. Research into learners with ADD, for example, shows that the effects of having ADD are often less harmful than the effects caused by growing up in an environment that is designed for the normative child. In other words, the condition of ADD is not the problem. The systemic ignorance, intolerance, and marginalization of the other is the problem.[13] Looking at ADD from a systems thinking perspective causes us to rethink the distinctions we make. For example, experts today are beginning to call spectrum disorders like ADD non-neurotypical, or NNT. While this new term still defines a NNT person from the perspective of a neuro-typical (NT) person, the term "typical" is less offensive than "deficit disorder," and much more accurate. Most of us would rather be atypical than disordered and deficient. These everyday distinctions can fall on a continuum from benign to moderately offensive to violent.

Marginalization of the other is the basis for us/them distinctions. Every thought-act or speech-act is capable of being a violent distinction. We must try to gain meta-level awareness of the distinctions we are making at each moment.

THE ETHICAL STANCE UNDERLYING SEEING PART-WHOLE SYSTEMS

Developing an awareness of the part-whole structure of all systems (and ideas about systems) allows us to see more, understand better, and develop a deeper sense of our place within the systems in which we exist. It is both quintessentially human and deeply ethical to know that we are part of something larger than ourselves.

We already illustrated the power of seeing parts and wholes with the firetruck example. In general, when we know to ask ourselves about the part-whole structure of anything (an idea, a system, a thing, a behavior, a feeling, etc.), we come to understand it more robustly. We see more parts. And when we see more parts, we provide a broader context for things and we can think more deeply about the interrelationships between and among the parts, including for human groupings.

[13] See Dr. Edward Hallowell's work on ADHD: Hallowell, N. (2009). Retrieved from http://www.drhallowell.com/meet-dr-hallowell/

Often when we seek to understand the distinctions we make, we must dig more deeply into the things we are distinguishing. We identify the parts of each idea to more accurately see if the distinction holds up under scrutiny. This is not only a powerful Apollonian skill, but also a Dionysian skill, so it develops the balance that characterizes the Odyssean mindset.

For example, in math we learn the distinction between an equation and an expression by seeing the equal sign as the part of an equation that differs from an expression. In science we deconstruct atoms, molecules, and organisms into their parts in order to make increasingly fine-grained distinctions.

In a different realm, we often feel complex emotions that are composites of several feelings, not just one. Being able to parse these complexities gives us clues about ourselves and insight into how to regulate or manage our responses to others and the world around us. Seeking deeper understanding of things and contextualizing them in nested whole systems is an important ethical skill.

At the same time that breaking things into parts (i.e., splitting or deconstructing) helps us better understand them, lumping things into wholes is an important skill. When people are different from us, lumping into wholes allows us to see our similarities. We can argue all day long about apples and oranges, or we can see that both are fruits.

It always seems a little strange when people say, "What is the black perspective on affirmative action?" or "What is the Latino perspective on immigration reform?" or "What is the Islamic perspective on terrorism?" Such questions presuppose each person (the parts) of the whole group shares precisely the same view. But this simply isn't reality. The truth is, while we can perhaps think about the Muslim perspective on terrorism, we should be willing to break this single perspective into a whole containing parts, which are sub-perspectives of the whole. Part-whole thinking can be conducive to both justice and fairness.

BELONGING AND PARTHOOD

There is an old saying, "No man is an island, entire of himself," based on the idea that we are inextricably linked in an infinite hierarchy of networks of interaction. What we call an "ecological ethos" means a mindset that we are all inextricably interconnected and nested within systems. The developmental psychologist Urie Bronfenbrenner held this ecological ethic. He developed an ecological model of human development that became the human ecology movement and led to the Head Start Program. The idea behind his theory was that"... interpersonal relationships, even [at] the smallest level of the parent-child relationship, did not exist in a social vacuum but were embedded in the larger social systems of community, society, economics and politics."[14] Figure 10.3 illustrates how

[14] Bronfenbrenner, U. (1979). *The ecology of human development: Experiments by nature and design.* Cambridge, Massachusetts: Harvard University Press.

Bronfenbrenner used an ecological ethic of (1) situating the individual in the wider context that he/she is a part of, and (2) seeking to understand the interconnectedness of this context. Simply put, Bronfenbrenner put the development of the child in a wider context of concentric part-whole circles and looked at the salient relationships across these systems, thereby revolutionizing the field of human development.

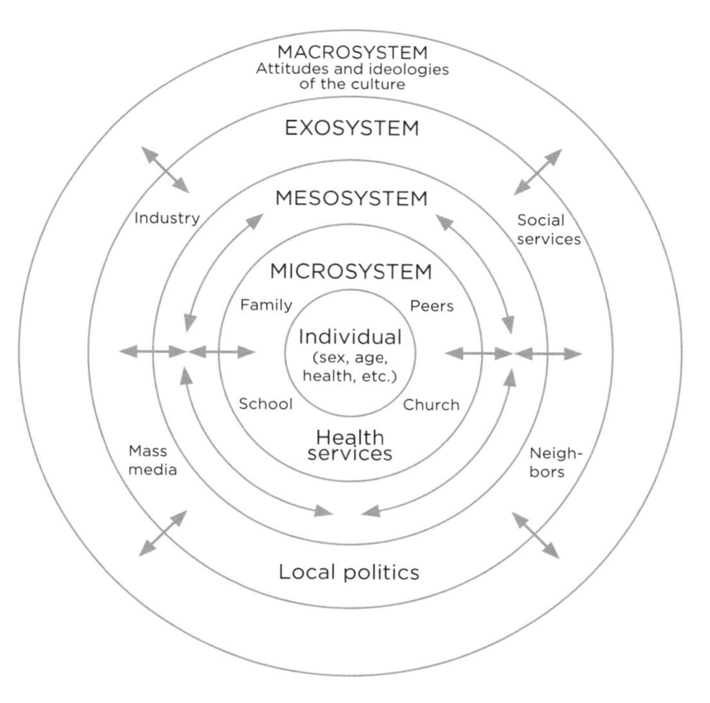

Figure 10.3 Bronfenbrenner's Ecology of Human Development

An ecological ethos is grounded in a combination of the systems rule (part-whole) and relational thinking (action-reaction). The relationship between the parts inside a whole is "parthood", alternatively expressed as brotherhood, sisterhood, partnership, and the like. The relationship between the parts and the whole is "belonging." Humans need to be a part of things. We all need to belong.

THE ETHICAL STANCE UNDERLYING RECOGNIZING RELATIONSHIPS

Action-reaction is not merely important in physical systems (e.g., Newton's third law), it is an essential metacognitive trait for understanding the interdependence of human relations as well as the interdependence of thoughts and feelings (cognition/emotion). Making action-reaction relationships a part of your everyday awareness means you understand that your actions have effects. Together, relationships and part-whole thinking create an ecological way of thinking in which we realize that we are part of something larger than ourselves and that hurting ourselves or others in the myriad ways we do only comes back to hurt the whole, of which we are but one part.

IDENTITY IS RELATIONAL

There's a tribe in Northern Natal in South Africa that uses the greeting "Sawubona," which literally translates as "I see you." Once seen, the person being greeted responds, "Sikhona," or "I am here." The exchange is borne of the spirit of ubuntu, a word that stems from the folk saying "A person is a person because of other people."

In other words, our identity is based on recognition by another that we exist. We are all defined by how we relate to others in context. At home, we are parents and spouses to those around us; at work we are CEOs or assistants. This is in contrast to our typical notion that our identities come from within; in fact, we get our identities from those around us (others). Thing and other are two sides of a single coin. There is no thing without an other.

To understand the thing-other relationship is also to appreciate the interconnectedness that we share with the other. Our relationship to others is the essence of our humanity.

The quality of relationships we choose determines our happiness and success. So many of us choose control over connection because we seek the immediate gratification that comes with control and fail to see the long-term gain that comes with connection. An ecological ethic, which understands the interwoven nature of connections and the webs of causality that result, provides us with a clear-headed view of the full range of consequences of our choices.

Understanding things less in terms of direct causality and more in terms of webs of causality is another ethical aspect of understanding relationships. Causes often come in bunches. When we understand the world as being the result of systems of relationships, we better approximate reality. This includes interpersonal dynamics and the situations in which we find ourselves and others in life. Webs of causality inform both the affective and analytical domains. Knowing to look for webs of causality rather than direct causality leads to greater compassion and less blame.[15] Analytically, we are also able to find optimal solutions because the real world works in webs of causality.

CAUSE AND EFFECT ARE NOT NEIGHBORS ON A TIMELINE

When faced with a problem, we often believe it "just happened." Often, it is a consequence of innumerable behavioral choices made over time. We get lung cancer because we smoked everyday. We get divorced because we didn't see the patterns in our relationship with our spouse. Seldom do things happen suddenly. The events in our lives are the result of webs of causality, not merely the most recent cause. Systems thinkers sometimes say that "cause and effect are not neighbors on a timeline." This idea tells us to look deeper and in many cases further back to figure out how we got here. Kabbalists explain it this way.

[15] This is a deep thought that distinguishes between linear and nonlinear (webs of) causality. Linear causality (e.g., X causes Y) allows for the concept of blame, focuses only on the cause that immediately precedes the effect, and is the basis for notions like free will, guilt, and punishment. Adopting a web-of-causality ethic means that one understands that an effect has multiple causes and that every cause is also being affected in some way by other causes. For example, if a boy steals bread we see beyond the simple blame and punishment and perhaps look at the wider context which is full of causality.

> *Our five senses are designed to perceive effects but not the cause. We see branches, but we fail to see the seed. Did you ever see a tree suddenly spring up from the ground for no reason–suddenly, there is a tree? Of course not. Every tree, has a seed that preceded it. Seeds however, are concealed in the ground. Our senses never detect the seed level. Unfortunately, do you know how we have been conditioned to look at life? Suddenly, there is a problem in my business. Suddenly, there is an illness. Suddenly, I met the greatest woman. Suddenly, business is booming. Suddenly, we are short of cash. Suddenly, there is a problem in my marriage. Suddenly, he had a heart attack. Suddenly, they found a lump. Suddenly, she woke up one morning and realized she wasn't happy. Is there really such a thing as sudden? Not really. There is always a concealed cause, a seed that precedes any "sudden" event.*[16]

THE ETHICAL STANCE UNDERLYING PERSPECTIVE-TAKING

Everything you think and every time you think, you are projecting your view onto the world. And while there is nothing wrong with sharing your views, we have an ethical imperative to be aware that our views are just that—views—and that they affect the way we see others. The value of taking a perspective and being aware of the perspectives one takes (and perhaps more importantly, does not take—which is an interaction of distinctions and perspectives) is enhanced clarity. When we shift perspective, we also transform the distinctions, relationships, and systems that we see and do not see. Being more aware of the consequences of perspective-taking reinforces our ethical compass.

While distinction-making causes us to see the other, perspective encourages us to take a walk in others' shoes and see the world from their vantage (or disadvantage).

There is much talk about empathy and compassion. The Dalai Lama explains that compassion is at the root of all things. Without the ability to take on others' perspectives there would be little empathy and compassion. The ability to listen, which is so essential for human communication and social interaction, would be, and unfortunately is, a meaningless exercise without perspective-taking.

When aware of perspective-taking we come to realize that most of the perspectives we adopt are actually someone else's perspective. Learning to truly see the world through another's point of view, without projecting our own bias onto it, is at the core of being prosocial, emotionally intelligent, and compassionate. A moral code is a set of values, and a value of any kind, by definition, contains bias because it is based upon a perspective that values one thing over others. Of course, the problem with moral values is that everyone has them and they can all be so different.

[16] Berg, Y. (2008). *The power of kabbalah for teens: Technology for the soul* (p. 14). New York: Kabbalah Pub.; Berg, R. (n.d.). Suddenly Syndrome. Retrieved May 17, 2015, from http://kabbalah.com/

Even the golden rule can lead to conflict because it dictates that we do to others as we would have them do unto us. However, "others" are not like us—they believe different things. Something that is polite in one country or culture might be rude in another. The golden rule only works if you're dealing with a homogeneous other who is more similar to you than different. We need instead to treat people in a manner analogous to how we want to be treated.

PITFALLS OF PERSPECTIVE

Lock-in[17] is a term that describes how systems self-organize toward certain ends. Once these ends are established, it can be difficult to reorganize them. Examples of lock-in are the QWERTY keyboard, the common clock, Beta vs. VHS, and the internal combustion engine. Each of these items came about because of many complex factors having little to do with which design was best. Lock-in often prohibits a system from moving toward innovation and efficiency.

Lock-in is a kind of systemic bias. We get locked in to our ways of seeing the world. Our worldview, mindset, biases, beliefs, strong feelings, and allegiances are all forms of lock-in that cause us to limit the number of perspectives we take and lead to closed-mindedness.

[17] Waldrop, M. (1992). *Complexity: The emerging science at the edge of order and chaos* (pp. 36-37). New York: Simon & Schuster.

FOLLOW THE RULES

The systems thinking we all need and want is by its very nature a complex, robust, and adaptive activity—an emergent property of simple rules. So to become a systems thinker one must merely make a habit out of using the DSRP simple rules. From that simple act, systems thinkers and systemic thought will emerge. There's an expression in the CAS world that "order comes for free,"[18] which means that when you focus on the simple rules of a CAS you'll get a lot more than you bargained for. In short, you'll get the Odyssean thinker that everyone wants.

Lazlo Bock, Google's Head of People Operations, revealed in his recent book *Work Rules!*[19] that the nation's number one place to work where nearly 2 million people apply for work each year was wrong on their original hiring strategy. Initially Google wanted Ivy League, Stanford, and MIT graduates with top grades and even included difficult brainteasers in the interview process. Today, Bock says Google's new hiring strategy doesn't care about grades, brainteasers, or what school you went to. In fact, they don't even care about whether the applicants know how to do the job for which they're applying. Bock explains that successful new hires

[18] Kauffman, S. (1995). *At Home in the Universe: The Search for Laws of Self-Organization and Complexity.* New York, New York: Oxford University Press.

[19] Bock, L. (2015). *Work Rules!: Insights from Inside Google That Will Transform How You Live and Lead.* New York, New York: Grand Central Publishing.

figure it out if they have a mix of problem solving skills, leadership, and diversity of thought.

Initially Google wanted Ivy League, Stanford, and MIT graduates with top grades and even included difficult brainteasers in the interview process. Today, Bock says Google's new hiring strategy doesn't care about grades, brainteasers, or what school you went to.

When we focus our efforts on the DSRP rules we get a lot for free. The resultant systems thinkers will be more aware of both their thoughts and their feelings, and more importantly, know the difference between them. Recent research shows that the ability to identify and then also differentiate emotions in stressful moments helps people access the information needed to solve their problems and make better informed decisions.[20] In other words, the application of cognitive ability to one's emotions allows people to better respond to difficult times, and therefore exhibit less self-destructive behavior and experience less anxiety and depression over their lifetime. In the long run, a person who develops emotional capabilities through robust, adaptive cognitive abilities like systems thinking will be a better citizen, a better parent, a better teacher, and contribute more to our society generally.

[20] Kashdan, et al. (2014). Unpacking Emotion Differentiation: Transforming Unpleasant Experience by Perceiving Distinctions in Negativity. *Association for Psychological Science*, Vol. 24 (1), 10-16

Teachable moments arise every day. As designers of humans, we need to program them with a powerful model of reflection that will bring meaning-making processes into consciousness. When we teach how to fish, rather than give fish, humans learn from everything they do. This is what leads to Odyssean humans, the kind of humans we all want to be and we all want our kids, employees, neighbors, bosses, and politicians to be.

These four simple rules of systems thinking offer awareness of our cognitive processes (metacognition), which can be applied to our full selves. Metacognition is the single greatest gift we can give to anyone, because it leads to an ability to understand and learn anything, and it lies at the root of human success in all endeavors. This holds true for individual humans, and for organized groups of humans, as discussed in the next chapter.

CHAPTER 11

CAS ORGANIZATIONS: SYSTEMS LEADERS WANTED

CAS ORGANIZATIONS: SYSTEMS LEADERS WANTED

Following DSRP simple rules creates systems thinkers who are lifelong problem solvers and learners who can provide the foundation for a learning organization capable of adaptive behavior. This chapter provides a completely new basis for organizational design and leadership called VMCL,[1] which is based on four core principles:

- Vision (V): a concise future state or goal;
- Mission (M): simple repeatable rules that lead to the vision;
- Culture (C): shared mental models that support the mission→vision;[2] and
- Learning (L): incremental improvement of the culture and mission→vision through systems thinking.

[1] In true fidelity to the originating VMCL theory, "C" refers to capacity and culture is considered a part of capacity (and the other letters are unchanged). According to VMCL theory, organizational success is dependent upon building various capacital systems: from engineering, sales, marketing, hiring talent, and technology systems to processes that govern meetings or ordering office supplies. Thus, in VMCL capacity literally means, "build the capacital systems needed to effectively do your mission every day." In this book, we focus on the most important capacital system—building culture—because an organizational leader who is learning systems thinking must prioritize developing a shared culture around the organization's vision and mission. In our experience working with numerous types of organizations, we have found that starting with culture prevents the aspiring systems leader from falling back into old habits of command and control structures, effectively negating the benefits of mission→vision and learning. For those new to systems leadership, the best approach is to think of the "C" in VMCL as culture in the beginning and then move to more advanced stages of designing and evolving a host of systems that enhance capacity, thereby thinking of the "C" as capacity in a more advanced stage.

[2] We use the "mission→vision" notation to signify that we are referring to a particular type of vision (a concise future state or goal), mission (simple repeatable rules), and the CAS relationship between them (that the simple rules of mission lead to the emergent property of vision). This relationship is signified by the arrow (→).

If DSRP are the simple rules individuals follow to become systems thinkers, then VMCL are the simple rules a group of these systems thinkers follows to bring about the vision of the organization.

In complex systems like human organizations, the way the system behaves is emergent (a product of varied actions by autonomous agents), which means that the results you want are emergent. Therefore, it is important for organizational leaders to understand that the results are not things to be developed directly, but indirectly by inculcating a set of simple rules in each group member.

We often hear myriad ways CEOs, administrators, and other organizational leaders struggle to get more out of their teams and their organizations in hopes of increasing both their internal efficiency and their external impact. They want to work smarter and make their mark on the world faster. The problem is usually that these leaders are working in their circle of concern trying to command and to control for complex emergent outcomes rather than working in their circle of influence with simple rules. Often these leaders have the mental model that more command and control will get more results. Rarely is this the case and certainly not in the long term.

Every human organization is, by definition, a complex adaptive system (CAS). As a result—and contrary to popular belief—all organizations are learning organizations. The task is not how to create a learning organization, as that is already built into our evolution as social animals. Instead, the task is to create an organization that learns how to survive and thrive through adaptation—an organization that learns every day how to do its mission→vision better, faster, and cheaper. In this modern age you want your organization to operate more like a complex living organism than like a complicated machine. You want it to adapt quickly to the changing environment, be resilient when times are tough, and most of all be dynamic and alive, because you need it to attract other living things, most importantly human talent. We now have the ability to purposefully design what nature has successfully evolved.

Applying nature's secrets about how complex systems work is worth the effort in today's ever-changing, competitive, complex world. Nothing on the planet beats nature for its resolution of paradoxes, elegance in design, and sheer creativity and genius. So when you want form and function, alignment between everyday effort and ultimate goals, and adaptation, look to the advanced sciences of complexity.

Derek gave a talk not too long ago at a business conference at Cornell University. It was a crowd full of MBAs, entrepreneurs, and business types. They invited him to speak because our models are used in business from Wall Street to the Silicon Valley. But Derek's actual expertise is in the analysis of complex systems, systems thinking, evolutionary biology, and nonlinear dynamics. One of the audience members at an open mic asked the panel to share the business book they would read if they could only read one book. Panelists listed the usual suspects: *Built to Last, Made to Stick, Behind the Cloud.*

At his turn, Derek said, "if you want to know how to build a successful organization, you should consult the experts and read Darwin's *Origin of Species*[3] or some other biology or ecology book about the fundamentals of organization, complexity, adaptation, and evolution." Unexpectedly, the crowd roared and began applauding and his twitter hashtag lit up.

Business books are almost always wrong because as soon as we see someone succeed we try to mimic it. Mimicry is a good bet, by the way. But when we codify that success as the rule set for all future successes, it inevitably fails because people, organizations, markets, industries, customers, and entrepreneurs adapt and evolve. They come up with new ways of doing things. Let us give you an example. Today, we all look at Google to see what they did so that we can make our organization the next Google. Google does OKRs (Objectives and Key Results), so we all need to do OKRs. The truth is, the next Google will do it differently.

3 Darwin, C., & Beer, G. (1996). *The origin of species*. Oxford: Oxford University Press.

In this modern age you want your organization to operate more like a complex living organism than like a complicated machine. You want it to adapt quickly to the changing environment, be resilient when times are tough, and most of all be dynamic and alive, because you need it to attract other living things, most importantly human talent.

When we want to understand how organizations work and how they can work better we should consult the expert and inventor of organization itself: nature. There are plenty of experts to consult: beehives, ant colonies, schools of fish, ecosystems, even human networks. Contrary to popular belief, humans are every bit as much a part of nature as any other species. We build big skyscrapers. So do termites. We have sophisticated communications. So do Orca whales. Granted, we are pretty darn good at a lot of things, but one area in which we could really learn more from the experts is in how we organize ourselves. A key lesson from nature is that your organization is a complex adaptive system. The degree to which you lead and manage it as such will largely determine its success or failure.

THE RIGHT MENTAL MODEL OF ORGANIZATIONS

A superorganism is a bunch of independent organisms acting in unison. That is precisely what every CEO and president wants their organization to be, even if they don't know it yet.

As it turns out, surperorganisms are remarkable because they have a knack for taking a bunch of self-interested, autonomous agents (say employees) and transforming them into much more than a well-oiled machine. You get an adaptive, organic, social, intelligent machine that is capable of many more things than are the individual actors alone, and without the often slow and blunt instrument of leadership.

If you think about it, this is why we start organizations in the first place, right? You wouldn't start a company to do something you could do on your own (plant a garden, build a dog house, etc.). Working with other people is a pain in the neck. If you could do it alone, you would. The very reason we start organizations is because we can't do what we want to do alone. That means something pretty important. It means that the whole purpose of any organization is to achieve collective action. As such, we should consult the expert on collective behavior: nature.

Instead, we do what we have always done. Let us illustrate. When we start an organization, we have the pesky problem of individuals with different opinions, experiences, lifestyles, and preferences. Getting all those people on the same page takes leadership. It takes a plan. It takes policies and processes. It takes command and control structures.

Actually, it doesn't. What it takes is something simpler. If you want a superorganism, your maxim should be to distrust what is complicated and trust simplicity. Simplicity will get you complexity, and complexity is what you want because complexity drives adaptation, robustness, and intelligence.

You might be thinking, "This sounds interesting but is it right for my organization?" It is right for your organization because your organization is a complex adaptive system (CAS). You don't get to choose whether or not you have a complex adaptive system for an organization. Because it's made up of humans interacting socially, it already is a CAS. You get to decide whether or not you embrace this reality and leverage it to your advantage.

If you ever go to India, Asia, or many places in Africa or South America, get yourself a coffee, tea, chai, or whatever, sit near a busy intersection, and just watch the traffic patterns. What you'll see if you take the time to notice is what looks like total chaos, but there is order in that chaos. Chaos means that a system is right on the verge of order, just before becoming random or stochastic (i.e., no order). So there is order, it's just less obvious. When you sit at the disorderly intersection with no traffic lights, vehicles of all sizes and speeds, you will perhaps begin to notice some things.

Figure 11.1: Video of CAS Traffic

crlab.us/stms

First, there is variability. The sizes of vehicles range from pedestrians (even small animals like chickens), to cart pushers, to donkey and cart, mopeds, motorcycles, tuk-tuks, small cars, minivans, SUVs, small trucks, buses, large trucks, and large buses. There's a diversity of speed, too, from old women slowly making their way through what seems to be insurmountable traffic, to zooming mopeds carrying dad, mom, and three children, and a wire cage of chickens; and there's a fruit and vegetable cart attached to a pedal bike, a minivan packed with tourists, and in the middle of it all is a big bus stopped in the middle of the road. Cars swerve and zoom in every possible direction. The sheer diversity of size, speed, and direction is amazing.

This diversity is in stark contrast to the straight freeways frequently found in the West where all the cars are heading in the same direction at any given time. All are traveling at a homogeneous rate of speed (somewhere near the speed limit). Most fit within a narrow range in size (from mini to SUV).[4]

Second, returning to our chaotic intersection, there are simple rules to the system, you just have to pay attention to find them. Back in the Far East, a couple of tourists catch your eye. They approach the edge of the sidewalk, wide-eyed and concerned. You can almost hear them asking themselves, "How do I get from here to there without being killed!?" Inside, they're thinking like a Westerner thinks:

This is dangerous;
The less time I spend in the danger zone, the higher the probability of living; and
Run!

What follows next is at best a lot of beeping, yelling, and general confusion caused by our Western friends disrupting the system. Their thinking makes sense. It is rational. It is logical. But it is also wrong. It is wrong because their mental models do not match reality.

If they had paid closer attention to the system they would have noticed the impossibly old woman, head down, her frail body taking tiny steps beneath the large brim of a wicker sun hat, cane wobbling. She walked across in a straight line. No dodging, jutting or serpentining. Straight. Slo-o-o-w. Steady. Her mental model of the system perfectly matched the system.

What does this old woman know that our smart Western friends do not? She understands the simple rules of the complex adaptive system. She understands that the simple rule is to avoid collision and not to go too fast to allow for adjustment of reaction time. She also knows that larger things have the right of way. What this means is that going slowly across the road is the safest way to proceed. Going quickly is the most dangerous![5]

4 Schulz, M. (2006, November 16). Controlled Chaos: European Cities Do Away with Traffic Signs. Retrieved April 17, 2015, from http://www.spiegel.de/international/spiegel/controlled-chaos-european-cities-do-away-with-traffic-signs-a-448747.html

5 Hans Monderman, a Dutch traffic engineer, focused on similarly "disorganized" intersections (those devoid of traffic signs and markings) and argued they were counterintuitively safer. These stripped-down intersections represent a complex adaptive system in that the agents (drivers) change their behavior (e.g., make eye contact with fellow motorists and pedestrians), producing a spontaneous, bottom-up type of order (Project for Public Spaces. Retrieved April 18, 2015, from http://www.pps.org/reference/hans-monderman/).

We see wrong mental models a lot in organizations. Because organizational leaders and members don't take the time to understand their organization as a complex adaptive system, they don't understand how to ensure they achieve their goal or desired effect on the world. Without fail, the organizational problems we face result from the difference between how organizations actually work and our mental models of how they work.

VMCL

Much like you would design an iPhone, you can also design an organization. So we look for the design principles of an adaptive, modern organization—an organization that uses simple rules (a mission) to bring about a goal (a vision). From our understanding of CAS, we know that if we want to influence the emergent properties of a system, we need to tweak the agents and/or the simple rules. It is through the agent-to-agent interactions that are based on the simple rules that the behavior of the system emerges. If we want to create the behavior of the system, we can't focus on the results, we must focus on the underlying rules that bring about the system behavior. That might seem really obvious advice, but you'd be surprised how seldom we follow it.

When we want an intelligent ant colony, which is comprised of a bunch of ants with single-neuron brains, we think let's start an ant training school to get the ants to be smarter. But the truth is, each individual ant is dumb. Ants will always be dumb. So how do we "make" them collectively smart? Simple rules. Place three piles of food at increasing distances around an ant hill and watch what happens. The closest pile gets carried back to the hill first. The second closest goes second. The pile that is furthest away goes third. How are dumb ants all of the sudden acting in intelligent ways that both optimize efficiency and decrease risk of exposure to predation? The answer is that they follow simple rules: (1) look for food, (2) if you find food shoot pheromones out of your butt, and (3) never cross a pheromone trail. These rules, on initial inspection, don't scream intelligence. But intelligence is what emerges out of this banal collective behavior. Sometimes the output you want (an emergent property) isn't attained through simply reverse engineering an analogous input, but through simple rules followed over and over again by individuals. When designing an adaptive organization, we must attend to the vision (V), mission (M), culture (C) and capacity to learn (L) of our system.

> **The organizational problems we face result from the difference between how organizations actually work and our mental models of how they work.**

Remember back in Chapter 2 when we discussed the learning feedback loop (Figure 2.7) and evolution of some complex adaptive systems on Earth (Figure 2.9)? You'll remember too that the image in Figure 2.7 we converted to a tiny inline image called a SparkMap ( ) because it represents the process of mental model building or systems thinking that constitutes learning. This SparkMap is so important we'll want to refer to it often. Figure 2.9 showed that Gell-Mann puts thinking and learning as the driver of cultural and organizational evolution—the ability to adapt. He contends that human evolution itself relies on the transmission of learned information among individuals from generation to generation. Individual learning and thinking underpin the evolution of individuals locally and the evolution of organizations globally. Learning ( ) is as important to the development of the individual as it is to the evolution of whole organizations. Learning ( ) drives culture, because culture is simply mental models ( ) that have come to be shared.

Learning (L) is the first part of the VMCL model because it is the thing that drives everything else. And we learned in Chapter 5 that DSRP (systems thinking) is the process by which information is structured to become mental models. Changing mental models over time is learning. So using DSRP to structure information to adapt our mental models is learning.

You will learn later that culture (C) is the process of a group of people learning in such a way as to share important mental models. So culture is actually caused when individuals in an organization begin to share the same mental models. In and of itself, the process of learning that leads to culture (which we will sometimes denote as learning→culture to indicate this relationship) can transform an organization. But learning and culture could go in many directions. To create an organization that operates as a superorganism, it needs a simple set of rules to follow (a mission (M)) and a goal state (a vision (V)). Now we must put it all together: [learning → culture]→[mission→vision] or VMCL. In Figure 11.2 you can see that VMCL is a cycle that revolves around building an organizational culture that supports the learning, mission, and vision. Figure 11.2 shows this cycle.

Figure 11.2 The VMCL Cycle

VMCL shows us that the most important leverage points in your organization are four-fold:

1. The selection of agents (i.e., hiring, training, motivating, and incentivizing talent to be systems thinkers and drive learning);
2. The explication of simple rules (your mission);
3. The explication of the goal (your vision); and
4. Building a culture that supports learning, mission, and vision.

Learning, mission, and vision are a function of culture. So the single thing that will bring these three things about is *culture*. Therefore, you need to do just one thing: *build a culture of systems thinkers that are laser focused on your vision, and the simple rules of your mission.*

How do you select your agents? Well, you either find them or create them. You either find people who are already systems thinkers or you get some people with the requisite raw intelligence and you train them. You build an environment that incentivizes them to select you, be drawn to your organization, and want to work there. The majority of this book provides examples of who you're looking for, or how to train raw talent to become what you need: systems thinkers *á la* DSRP.

Once you have your agents, you need to build culture. Culture is comprised of shared mental models. The more mental models people share and the more deeply they share them, the more robust the organizational culture. So to create culture, repeatedly build and share mental models among all of your agents. Therefore to build culture, build and share mental models.

The most important of all the mental models comprising an organization's culture is its vision and the simple rules of its mission. Vision is the future change we want to see and mission is what we do every day to get there. We will soon explore how to create a CAS version of a mission→vision that completely differs from what you have been exposed to in those boring offsites.

You need to do just one thing: build a culture of systems thinkers that are laser focused on your vision, and the simple rules of your mission.

Once you have your vision and mission enculturated as a shared mental model across your organization, the single job your people have to do is use systems thinking to learn how to do their mission faster, better, and cheaper to more effectively and efficiently achieve your vision.

To be visionary means that we see into a particular future and lead people to see and work toward that future alongside us.

To have a mission means that we work doggedly in the present to build our vision. The mission of an organization is the simple rules of the organization. The simple rules of your mission, done repeatedly by many, will bring about your vision.

Finally, there's people. Independent agents. They are not human resources, they are human beings. They exist autonomously and independently from the organization. They are a thinkforce, not a workforce. Systems thinking helps them to constantly build mental models that better approximate reality. They constantly adapt and improve by subjecting mental models to feedback from the real world, with the very specific purpose of learning how to do the mission→vision better, faster, and cheaper.

It's simple, yet few organizational leaders know that's how to do it. They get stumped, often because their mental models of what their organization is and how it works are wrong. Also, our mental models of what a vision is, what it does, and how it differs from mission are way out of whack. Let's take a look at the way we think an organization works and then we'll take a look at how it actually works.

Culture is comprised of shared mental models. The more mental models people share and the more deeply they share them, the more robust the organizational culture. So to create culture, repeatedly build and share mental models among all of your agents.

THE TRADITIONAL MODEL OF ORGANIZATIONS

Remember that wicked problems result when how we think our organization works is out of alignment with how it actually works. The traditional model of organizations is based on several significant mental model errors. The most serious errors are highlighted in the next few pages, with a suggested alternative mental model to maximize your organization's potential.

Let's take a closer look at our most basic assumptions about how organizations work. We appear wed to a traditional yet erroneous mental model of the organization. Why are we committed to a model that is so often wrong? We begin by elaborating—only slightly facetiously—the traditional model of organizations. The traditional model can be summarized as plan, command, and control resources.

PLAN

First, we need a *plan*. We used to think we could plan 10 years in advance. Then someone figured out that was crazy talk and the 5-year strategic plan was born. Ironically, it took more than 5 years to realize that 5 years is also a really long time. Compounding this fact, the speed of change in markets, society, culture, and technology is accelerating, making each year an even longer time, relatively speaking. So we moved to the 2-year plan. The 2-year plan was much better because it was nearly 60% more accurate! Of course, we soon realized this too was a really long time in business years, which are roughly akin to dog years. So we've settled on "we just need a business plan because we can't get money without one." The basic idea though is that there is something magical about a plan. You have to have a plan… or do you?

We are not against planning. We are against hubris. Planning, as it is currently practiced, looks a lot like hubris. Hubris that you can predict the future, account for all the variables and all the actors in the complex system that makes up your business, your market, your industry, or the global economy.

The truth is, there's a lot of luck and randomness in complex systems. And there's a lot of complexity and interactions that cannot be known. Unless you have a crystal ball (or Google's server data), you're better off creating an adaptive culture with a solid mission→vision than trying to predict the future.

COMMAND

Second, we need a *command structure*. Everyone loves hierarchical trees, so let's use that. In fact, when we really want to understand an organization and how it works, simply ask to see their organizational chart or "org chart" (see Figure 11.3). That is the best indicator of how things work in an organization. Because everyone knows that if your boss asks you to do something you definitely do not want to do, or you disagree with, there is no way to manipulate or otherwise obfuscate in a way that effectively means it won't get done.

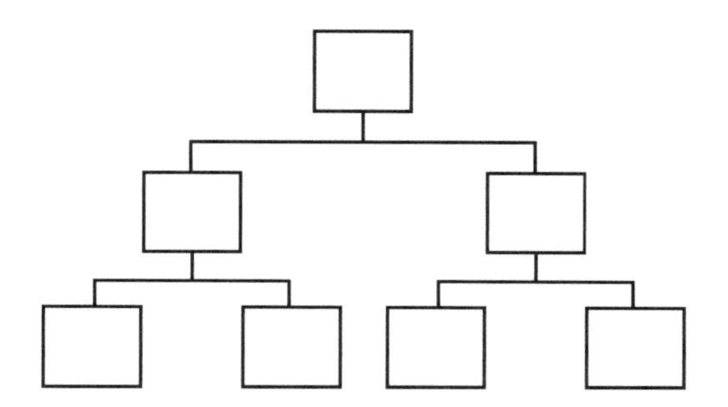

FIGURE 11.3: Hierarchical Org Chart

The org chart says who is who's boss, who reports to whom, who to see if you have a problem. The truth is, it's bunk. It's not how organizations work. Sure, bosses can delegate, hire, and fire. Sure, bosses can tell employees what to do and, to some degree, use hierarchical power to get them to do it. But it's also true that employees can stall, drag their feet, ignore, play dumb, play dead, go around, or raise Cain. Direct reports can and more often than not do make the ultimate decision on what they do. If they have to do something, they can decide to be strategically compliant rather than authentically engaged. They are in charge of how well it gets done. None of that is captured in an org chart. Yet this kind of stuff constitutes 90% of the workweek. Why not create a different mental model of an organization—one that better reflects reality?

An org chart tells us that our organization looks and acts like a command and control hierarchy, when in reality organizations look and act like dynamic social networks (Figure 11.4). In a social network, the nodes (things being connected) are people and the connections (lines) are the relationships between people. The people with the most and highest quality relationships are the most connected, and therefore the most influential in the system. Let's call this the relationship-value (or, R-value) of a node. Notably, the thing that we call command and control or hierarchical influence is in actuality the relational influence of the people in the network. So what we should be looking at in our organizations is people and their R-values, or the influence they have on the system.

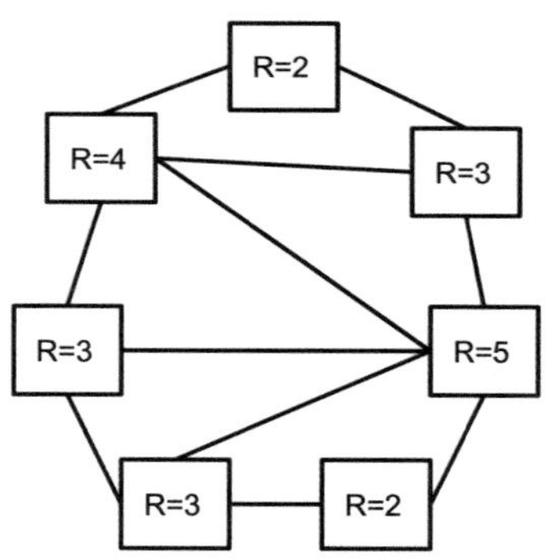

FIGURE 11.4: Social Influence Network

In healthy organizations, the people with the highest R-values are also the people located near the top of the hierarchy. The stated leaders are the actual leaders. For example, such a boss is at the top of the hierarchy because she has more influence, more knowledge, knows more people, has her hands in more processes, and is a part of more initiatives. She's more knowledgeable because she is, and has been, more connected.

That's not to say that command and control hierarchies don't or can't work. But if you want an adaptive organization, command and control doesn't work. The reason why in the simplest terms has to do with communications and time. Simply put, when something occurs on the ground, the employee needs to respond to it right then and there in order for the organization to be adaptive. In the case of the customer, this is critically important. But if the employee has no authority to think or is not incentivized or rewarded for thinking, then they have to go check with higher-ups, sometimes through several levels of command and control. The time it takes for the message to travel up from the employee in the field to the organization's "central nervous system" and then back again kills any hope of an adaptive, responsive interaction on the ground. We see this in complex systems in nature. There's simply not enough time to have leaders tell us what to do because leaders can't be everywhere at once. Or can they? What if everyone is given a leadership role in the organization to make decisions locally based on simple rules rather than hierarchical command and control?

The org chart mental model versus the social network mental model is important for other reasons. First, the org chart model assumes that people are extrinsically (externally) motivated (e.g., if my boss says to do something, then I should do it). What is more likely true is that people are semi-autonomous, somewhat selfish, independent social agents with their own set of extrinsic and intrinsic (internal) motivations. If what the boss is asking aligns with those motivations, then there is compliance, but if not, the behavior you'll see is false engagement such as strategic compliance, ritual compliance, retreatism, or rebellion.

An employee is strategically compliant when they do what you ask of them for their own strategic interests. This also means that as soon as their interests are out of alignment with yours, they will not comply. This demonstrates the problem with org charts—they give us the false sense of employee interests being aligned with organizational values. Ritual compliance is when a person complies because they are accustomed to doing so, i.e., out of habit. That might work for turn-of-the-century factory workers, but not for those working in today's knowledge economies. Today, many factory jobs require authentic engagement and thinking even on the assembly line because the tasks, although routine, are highly technical.

Retreatism occurs when a person disengages either physically or mentally, causing significant inactivity or reducing productivity. He shows up and collects a check. Rebellion against the instructions can be either active or passive. Active rebellion is more obvious, for example storming out of a meeting or refusing to do a task. Passive rebellion is more popular, such as resorting to subversive behavior that can be extremely damaging and costly to your organization. In short, the org chart fails to capture these details of how organizations (and the people that constitute them) actually work.

Never before has this been more true than in the current and dramatically changing business climate. Some will argue that command and control hierarchies have worked for centuries. And they would be right, sort of. In factory or industrial labor situations, command and control hierarchies look like they work because there is more strategic and ritual compliance based on power differentials and available choice. But today in many market sectors where people shop for workplaces like they shop for the best deals on consumer goods and where hiring the best talent is the crux of organizational success, the best employees have more choices. In addition, as we enter the knowledge age, workers are more educated, more savvy, and more empowered to act upon their own interests. This means that today's org chart mental model is more fiction than fact.

It's never easy to give up something that we are so accustomed to. Still doubtful? Still in love with org charts? Okay. Pop quiz.

Organizational change is: (choose one)

A. Easy

B. Hard

If you answered A, you either work in an organization comprised of 1 person, have never tried organizational change, or are an organizational guru and need not read any further. If, like most of us, you answered B, here's why: command hierarchies don't actually work as promised. If they did, organizational change would be easy. All you'd have to do is decide at the top how things should be and then pass it down the hierarchical chain. At each level of hierarchy each person would do exactly what their boss wanted and *kerpoof!* organizational change.

Anything more than superficial organizational change requires a shift in the culture of the organization and that requires individuals to change their mental models. It requires that individuals adopt a new mission→vision. An org chart simply cannot account for this type of change. But an MV graph can.

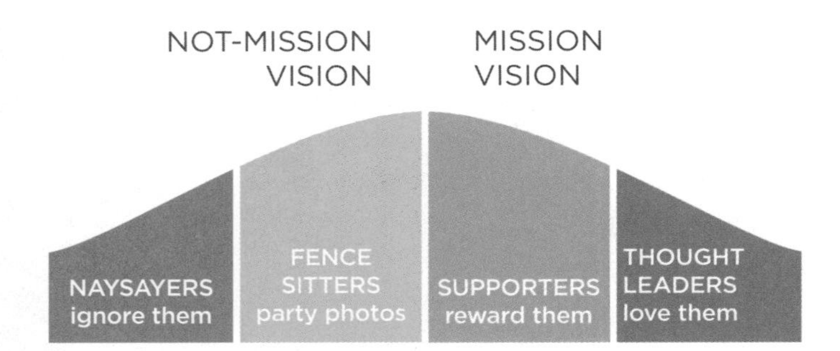

FIGURE 11.5: MV Graph For Culture-Building

Figure 11.5 shows a better model for organizational change—a mission→vision (MV) graph that determines who is on board and who is not (and to what degree). The MV graph is not a compliance model, it's a culture-building model. Fence-sitters need to be motivated to become part of the culture shift. We have found that MV graphs work very well with organizations that are struggling or in which leaders don't have the kind of absolute hiring and firing authority that CEOs might have (e.g., school districts, government bureaucracies). In such cases, MV graphs help identify work that needs to be done to make the shift toward a new organizational culture.

MV graphs can be used in the same way that a Senate Majority Leader might place images of senators in the Yay or Nay column in preparation for a vote. Placing individual employees on the MV Graph helps leaders know where the work needs to be done and how to adequately incentivize each individual. On the right side of the MV graph are systems thinkers, comprised of supporters and thought leaders. These folks support the mission→vision. On the left side of the graph are non-supporters: fence-sitters and naysayers. The idea is to manage the process of getting as many people as possible from the left to the right side of the graph, thereby creating a critical mass of support. Thought leaders in the organization typically don't need much more than camaraderie and appreciation to continue their work advocating the mission→vision. We generally give them love and support. The majority of those who buy into the mission→vision can be called supporters, and this is where your incentives and rewards should go to effect change. Organizational leaders make a mistake here by rewarding fence-sitters in order to get them to move to the right, but this actually has the opposite effect, which is to motivate fence-sitting. Fence-sitters are waiting to see what's going to happen so you want to avoid rewarding this behavior. But you also want to give them reason to become contributing systems thinkers. Do this by showing what we call "party photos," which are various communiqués designed to let fence-sitters know that being a part of the systems thinkers is the best place to be. They get rewards, are having fun, and love what they're doing. Resist giving fence-sitters anything but party photos. Finally, many leaders get sidetracked by getting into control battles with naysayers when the best strategy is to ignore them in a kind of Kung Fu-style deflection.

We don't expect you'll go in tomorrow and throw out your org chart, although that'd be a step in the right direction. Organizational leaders are far too in love with this dysfunctional mate to give up on him right away. At most, perhaps you'll see that behind the org chart is a mental model of how an organization works, and understand that maybe it doesn't work quite that way. Maybe it's not quite that clean. Maybe there's a more accurate mental model. That's the power of mental models. You are not the problem, it's the mental model that is the problem.

Continuing to believe that the org chart mental model is the mental model of your organization is a little like our Western tourists insisting that racing across that intersection mentioned earlier is the best way. Be wise like the old lady. Ensure that the mental model you have matches the organization you have. If not, it makes no difference how clean or beautiful your mental model is or how it makes you feel inside. If it is at odds with how your organization actually works, then it's wrong.

CONTROL

Next (in the traditional model), we need a *control structure*. We need processes because we need to make sure that we understand every single step of a process that has not happened yet and that will constantly be changing. The ideal structure for this doesn't exist, so we will repurpose our favorite structure of all time—the org chart. Simply turn it on its side and *Voila!* the flow chart. Excellent!

An org chart is really just a "tree" network. It follows a pretty typical pattern of branching where there is a trunk (inverted) at the top (CEO, Executive) that leads to the main branches (VPs, Directors, etc.), which in turn branch down into the next level of hierarchy and so on.

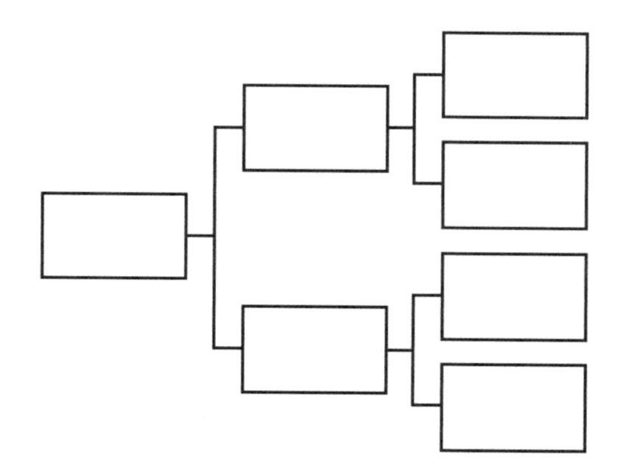

FIGURE 11.6: Flow Chart

We love trees because we love hierarchies. They make us feel safe. They make us feel like we are in control. They get us. We can't quit them. We love our trees so much that when it comes time to get down to action steps, we go straight to our "go to diagram"—the org chart.

Org charts make us feel in command of the troops and flow charts make us feel in control of processes. We like to feel in control, even if we are not. Here again, we feel the need to make it clear that we are not against org charts or flow charts. *Our concern is with mental models that are out of alignment with reality.* Our experience shows that org charts and flow charts are two of the most popular ways that we incorporate faulty mental models into our organizations. We used a flow chart last week, and not too long ago we used an org chart. The difference was, it was an appropriate use. In the case of the process, it was a very simple, linear process that followed four sequential steps. That's a perfect place for a tree-like flow chart.

So if tree-like flow charts are the norm, what is the alternative? Well, for simple linear processes, a linear process diagram is a perfectly fine choice. But if you want to understand how a bunch of people are somehow going to start acting like a superorganism to get things done in an adaptive way, then we need to revisit simple rules and complexity.

The best way to get a group of people to work together to bring out the power of their collective dynamics is to give them a simple set of rules and a shared vision (a "North Star"). Think of it this way. If you want a group of people to end up in a particular location, you have to tell them where it is, motivate them intrinsically and extrinsically to get there, and tell them the rules they should follow on the journey. What you don't want to do is micromanage the journey, because things on the ground might be difficult and complex and require adaptivity and grit. What you want to say is, here's the destination in all its glory and here are rules for getting there.

So let's say you know where you are and where you want to go (vision). We've been taught since third grade that the shortest path between two points is a straight line. And since third grade we've been imposing this limited truism like a bully on every process we meet.

Although it's mathematically true that the shortest path between two points is a straight line, it's not true in practice when we have countervailing forces and unknown or unexpected variables.

Let's say, for example, that in between point A and point B there is a swamp. Add alligators or botulism to the swamp and it makes it all the more necessary to go around. Suddenly the most costly, most time consuming path is the straight line. Yet this straight line or linear mental model is more popular than you might think. We use it whenever we *think* that the path between A and B is completely definable even though it is a path in the future through a complex and ever-changing landscape.

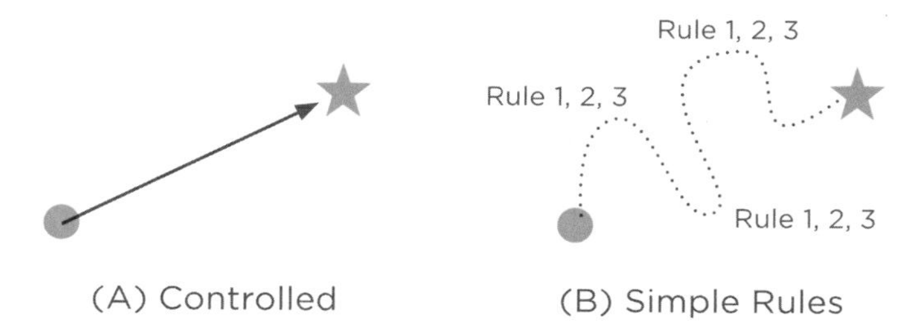

FIGURE 11.7: 3 Mental Models of Process

Figure 11.7 shows two different mental models. In (A) we say where we are, where we want to go and how we will get there. This represents linear thinking. In (B) we see where we are and where we want to get to, but our approach to how to get there is a bit different. Instead of a simplistic and linear projection, we take into account the unknown. This path is nonlinear, projected rather than definitive, and adaptive, using simple rules to guide what might be a complex journey.

MANAGE RESOURCES

Fourth (in the traditional model), because flow charts, org charts, and plans only work when we have enough human power, we're going to need some *human resources*. Also, humans inherently like to think of themselves as resources rather than individuals with their own agency and destiny. That's why they love to join a workforce managed by a Human Resources department. The idea here is to give the humans the plan, give them a flowchart and an org chart, tell them who they report to and then squeeze the human juice out of them to fuel the machine. It's all very mechanical, but it works, right? It does work, right?

Yes. It works. This traditional model of organizations does exactly what it is supposed to do. Exactly what it was designed to do: to make us *feel* in command and in control. It makes little difference to us whether this model of organizations is an accurate reflection of how organizations actually work. What's important is that it feels like it's working. This warm and fuzzy feeling of control is all we need, until we are faced with results we don't expect. It's these lackluster results that cause us to look for a different model. Ideally, that's when organizations realize their need of systems thinking.

Humans are quirky animals. We are independent and seek to differentiate ourselves and our identity and yet we are also extremely social animals who long to be part of a group. We have these amazing logical and analytical abilities (executive functions) and at the same time we can be petty, emotional, and retributive. We are both selfish and social.

Businesses of all kinds need resources: financial, physical, and human. We don't think of ourselves as resources. So it's a bit of a disconnect when organizations think of us this way. We have families, loved ones, personal goals, and aspirations. When these priorities, goals, and aspirations align with an organization' s, then it's a great fit. When they don't, it's a bad fit. Yet the primary mechanism for working with humans is a department we almost always call Human Resources. That's never a good department. We should get rid of it in exchange for Human Engagement or Human Talent or really anything else that demonstrates an understanding of the paradox that is humanity.

People want to improve. They want to be challenged to be better versions of themselves. They want to use their talents to make things better for themselves (selfish) and others (social). We humans like to be engaged in what we are doing. The best way to engage a human is to engage their thinking. Everything else is just strategic compliance. And as soon as the transactional exchange ends, the engagement ends. But engage people's thinking and you engage their human spirit. You don't get 40-hour a week employees, you get employees who think about things in the shower, on the way to work, and before they fall asleep.

Extrinsic rewards and incentives (e.g., "carrots") work because we are behavioral animals. We are like Pavlov's dog. But intrinsic motivation is rooted in engagement. Engagement is the result of three things: using your brain to solve problems, feeling like you're part of something larger than yourself, and effecting change.

The greatest threat to the traditional mental model of organizations is *results*. When we don't get the results we want, we are forced to choose: do I want to feel in control or do I want to get results?

Table 11.1 summarizes the mental models that influence traditional versus systems thinking leadership of organizations.

Table 11.1: Distinguishing Mental Models

Traditional Leadership	Systems Thinking Leadership
Over-engineered plans	A culture that constantly adapts and evolves its mental models
Command hierarchies (org charts)	Lead complex social networks, intrinsic and extrinsic incentives, MV graphs
Control processes (flow charts)	Face hard truths about where we are (realism in mental models), have clear vision of where we are going, and simple rules for how to get there (mission)
Utilize a workforce (manage human resources through command and control)	Empower a thinkforce or systems thinking force (inspire talent to constantly learn and improve)
Top-down management of outcomes focuses on circle of concern	Bottom-up management of outcomes focuses on circle of influence (i.e., focus on agents and simple rules and adapt to alter system behavior/outcomes)

VMCL is a systems leadership and organizational design model that helps us to better lead and manage organizations.

THE SUMMIT PARADOX

Prior to becoming a scientist, Derek was a mountain guide and high altitude climber. In some ways he was doing the science of observation all those years as a guide. He learned a lot about group dynamics, leadership, and human behavior in stressful situations by observing people on short, 7-day courses to longer courses lasting 90 days. He learned how people and groups acted in high-pressure situations. He also learned a lot about the universality of people because he led trips for corporate executives, Wall Street brokers, gang members, drug addicts, alcoholics, adjudicated youth, college graduates, farm kids, city folk, and entitled suburban "victims of wealth." All of this occurred in a context: the mountains, deserts, and canyons of the world (and on rivers while raft guiding). So while he watched the complexities of human dynamics play out before him, he also paid close attention to meteorological, ecological, and geological patterns in nature. In many respects, this was his scientific training, because he lived for decades in a lab where he could see human and natural systems play out. This training period has informed nearly all of his scientific research and entrepreneurial endeavors.

We tell you this because planning a high-altitude expedition has a lot in common with the relationship among vision, mission, and culture, and provides an anecdotal yet effective understanding of how the relationships between vision, mission, culture, and learning work.

When you plan a high altitude expedition the vision is to get to the summit. Its a simple vision—an image one burns into their mind of standing on the summit. Yet beyond those mental snapshots of a future moment, the vision is at once all encompassing and totally irrelevant—a paradox of sorts. It is the *raison d'etre* of the climb and yet it is also not something you can focus on if you ever hope to actually attain it. So you need something to do. For mountaineers, this is called the mountain rest step. The rest step is a way of orienting your body such that in between each step one straightens out their body and standing leg so that all the weight of you and your pack is being supported by an efficient skeletal system and not by muscle. For the brief moment while you rest in between steps (sometimes micro-seconds and sometimes minutes), the body isn't using its muscle to support your weight. This gives the leg muscles a micro-rest, thereby allowing the climber to take more and more steps. The primary function of the mountain rest step is to make it so you can repeat the action for, in many cases, what seems like a really long time. This is especially important as you gain altitude and the rarified air loses more and more of its oxygen content. As your brain has less and less oxygen, thinking becomes less clear, muscles tire more easily, and our remarkable human cognitive functions are reduced to a simple mantra: take one more step, repeat. Over time, although it sometimes may seem impossible, these repeated steps add up, and with a little luck you will reach the summit.

We can think of the summit as the vision. We can think of the mantra, "step, repeat" as the mission: a set of simple clear instructions on what must be done over and over again in order to attain your vision. The motivation to reach your vision and the stamina to carry out your mission day after day is a big part of the success of any expedition.

Visions are emergent properties of organizations, and are worked on and toward indirectly through missions. Let's take the summit example. If my vision is to stand on the summit of Mount Everest, I can't work on that vision directly. I don't do standing exercises or take photos of myself repeatedly standing on a balance ball with an ice axe and a flag. Even though my vision—to stand on Mount Everest—is my beacon, my North Star, the thing I'm aiming to achieve and the muse for all of my efforts, it turns out that it's not something I can work on directly. I can only work on other things that will indirectly get me there. The thing that is the *raison d'être* isn't the thing that I work on everyday. I have to work on other things in order to get there, indirectly.

Another important factor that determines the success or failure of an expedition is called "expeditionary behavior" (EB). EB is the culture of the expedition. It is the common set of mental models that members of the expedition share (or often times don't) that makes the expedition a success or failure. It is less important what the culture of the expedition is than that there is one, by which we mean there is a shared set of understandings about the purpose, rules, and norms of the expedition.

Finally, there is learning. Learning is the key to all expeditions and to culture itself because learning is how people adapt to situations that don't fit their mental models. Each climber spends a lifetime of learning prior to the climb and is learning throughout the climb and adapting their actions and behaviors to each new understanding.

Now we want you to look at this expedition from the perspective of time and resources and compare that to what you value. Climbers value the summit the most, yet by the time you reach the summit, you likely reflect on the fact that many accidents occur on the way down, you may be burning daylight needed to safely descend, and often you are suffering from a debilitating headache and nausea from the altitude, and exposure to cold or intense sun through decreased atmosphere. The result is that climbers often spend a very short period of time on the summit, sometimes just a few minutes, before heading down. The mission, climbing one step at a time, may last from a few days for smaller mountains to up to 30 days for larger ones. The value is high because, well, climbers like climbing. The culture, which can take weeks or months or sometimes years to build amongst climbers, is of relatively low perceived value. And learning, which can take a lifetime, is often not factored.

Table 11.2: The Summit Paradox: What We Spend Time on Versus What We Value

	Time/Resource Commitment	Perceived Value
Vision: Standing on the summit	Minutes	Extremely High
Mission:Climbing ("take one step, repeat")	Days/Weeks	High
Culture: Expedition behavior	Weeks/ Months/ Years	Medium
Learning: Climbing skills & training	Years/ Lifetime	Low

It is a paradox that we spend the least amount of time and resources on the things we care about the most.

Think about your own organization. You may value your vision and customers, but the amount of prep work that goes into developing the learning and capacity to execute your vision trumps the actual amount of time you spend pleasing customers.

Why does this paradox concern us? Well, if most of your time is spent doing things that are less valuable to you, then it becomes increasingly important to *align* the things you value but spend little time on with the things you spend the most time on. Focus is key. If you don't focus your learning and culture-building around your mission→vision, then you'll spend even less time doing and seeing the things you value most.

MISSION→VISION

Table 11.3 essentially summarizes the argument thus far. Pay particular attention to the bit about where you can be effective. Above the bold line you have little influence or control because the collective dynamics of the system are too complex. Below the line you have greater influence and control to select/train the agents or tweak the simple rules. What you can see in the column entitled Traditional Organizational Model is that we try to command and control at all levels using the flawed mental models discussed earlier. This is in direct contrast to the alternative model of an organization: a complex adaptive system model, which we discussed at the start of this book. The Systems Thinking Individual Model shows that when we focus on DSRP, the complex kind of thinking that we all seek will emerge. The agents in the case of the individual systems thinker are ideas. As we move to creating a systems thinking organization, the systems thinking individual becomes an agent that needs some simple rules to follow (mission) and the organization needs a goal (vision).

The only thing left is to solve one final problem: with near perfect consistency, corporate mission-vision statements suck. How do we make them not suck?

Table 11.3: Access Points for Effective Intervention

Where You Can Be Effective	Traditional Organizational Model	CAS Model (superorganism)	Systems Thinking Individual Model	Systems Thinking Organization Model
Above this line you have little influence or control, the dynamics are too complex	Strategic plans	Emergent Complexity	An adaptive systems thinker	An adaptive super-organization
	Command organizational hierarchies	Collective behavior & self-organization	Formation of systemic thought	Formation of culture
Below this line you have maximum influence and control to train the agents or tweak the simple rules	Control organizational processes	Local simple rules	DSRP (simple rules)	VMCL, sharing mental models, especially the mission (simple rules)
	Utilize human resources	Autonomous agents	Ideas/information as agents	Adaptive systems thinkers (as agents)

INEFFECTIVE MISSION-VISIONS

First it helps to understand what an ineffective mission-vision looks like. In simple terms, if a mission-vision isn't doing it's job, then it's ineffective. The job of a vision is to ensure that everyone knows the future goal the organization is trying to reach. The job of a mission is that everyone knows the simple rules for bringing about the vision. Also, a vision and mission are coworkers, so it's also their job to work well together. That's why we arrow-hyphenate mission→vision rather than saying vision and mission. If the vision and mission aren't performing these three collective job duties, then they are ineffective.

With near perfect consistency, corporate mission-vision statements suck.

In the same way that an organization can be successful with an ineffective CEO, it can also be successful with an ineffective mission-vision. But an organization with an effective mission→vision, like one with an effective CEO, will be all the more successful. One of the best ways to ensure that you don't end up with an ineffective mission-vision is to understand what they look like and then distinguish those from effective mission→visions. So that's what we'll do. Here's some ineffective mission-visions.

Despite our deep and lasting affection for the work of the Santa Fe Institute (SFI), it is terrifically ironic that the mission-vision is ineffective; one would expect them to know more about how to manage complex adaptive systems. And one can imagine how much more of an impact SFI might have with a truly effective mission→vision:

- **mission**: The Santa Fe Institute is a transdisciplinary research community that expands the boundaries of scientific understanding. Its aim is to discover, comprehend, and communicate the common fundamental principles in complex physical, computational, biological, and social systems that underlie many of the most profound problems facing science and society today.
- **vision**: Many of society's most pressing problems fall far from the confines of disciplinary research. Complex problems require novel ideas that result from thinking about non-equilibrium and highly connected complex adaptive systems. We are dedicated to developing advanced concepts and methods for these problems, and pursuing solutions at the interfaces between fields through wide-ranging collaborations, conversations, and educational programs. SFI combines expertise in quantitative theory and model building with a community and infrastructure able to support cutting-edge, distributed and team-based science. At the Santa Fe Institute, we are asking big questions that matter to science and society.[6]

---

[6] Santa Fe Institute mission and vision. (2015, January 1). Retrieved April 16, 2015, from http://www.santafe.edu/about/mission-and-vision/

This mission-vision is complicated and ineffective. It fails to provide a concise and clear future goal (vision), the simple repeatable rules to get there (mission), nor do the vision and mission play well together (they read like two disconnected descriptions). Perhaps worst of all is that if you asked anyone at SFI to recite their mission or vision, we suspect most could not. A mission or vision this long—which is the norm—is longer than most people are willing to commit to memory and too long to be shared as a meaningful mental model.

A vision and mission can be short and still be ineffectual. Here's another ineffective mission-vision, which you likely will concur is like so many such statements gracing corporate headquarters and websites around the world. Tragically, this one we found leaning against the wall in the corner of a bank. It was professionally printed on an expensive, framed, foam-core poster, a indication of its importance during what was likely its brief heyday. One can imagine the meeting that led to its creation, which undoubtedly was painfully long for its attendees, only to have their time and effort left unloved in the corner.

- **vision**: To be the trusted resource.
- **mission**: We are the trusted financial resource for our members, serving them as the financial institution of yesteryear, with all the conveniences, technology, and accessibility of tomorrow.

The problem with this vision and mission is that they defy logic, which is likely the result of not understanding the purpose of a vision and mission in the first place. If the future goal is to be the trusted resource and the mission explains that they already are, isn't it time to close up shop? Mission accomplished, er, vision accomplished?

To find ineffective visions and missions, one need not search far and wide. A visit to nearly any company website will yield a vision and mission that feels more like it was created because "we're all supposed to have one" than as an essential and useful foundation for the organization.

10 TESTS FOR EFFECTIVE MISSION→VISIONS

Learning to recognize what makes a mission-vision ineffective and what makes it effective is the first step to creating a great mission→vision for your organization. Here are two of the best mission→visions ever created:

MISSION: Convert the unconverted

VISION: A Catholic world

MISSION: Go forth and multiply

VISION: Resilient biodiversity

Figure 11.8: Two Excellent Mission→Visions

What would be helpful though is a litmus test that guides you to develop a great mission→vision. Such a test would make it clear why the mission→visions in Figure 11.8 have been so effective.

We developed 10 tests to help people create mission→visions that do not suck and to help systems leaders to implement VMCL. Follow these tests in the creation of your mission→vision and we guarantee that it will not only not-suck, it will become the tool that gets you what you want—an organization that acts like a living, breathing, adapting superorganism.

We've used these tests hundreds of times with clients and they rarely ever fail. When they do, it's usually because clients fail to do #9 and #10. These tests work for designing or redesigning your organization. They provide a lot of things, but one thing they do not do is implement themselves. If you're not willing to do the work, don't bother doing the thinking. Using this technique, we're always able to uncover a powerful mission→vision that thrills both executives and employees.

Here are the 10 tests. Tests #1-8 relate to vision, mission, or both, while #9 and #10 address culture and learning:

Test #1: Mission→visions are short and simple;
Test #2: Visions capture a picture of a binary future state;
Test #3: Visions are intrinsically motivating;
Test #4: Missions are simple rules that follow a formula;
Test #5: Repeatedly doing your mission should bring about the vision;
Test #6: Mission→visions must be measurable;
Test #7: Mission moments are rare and precious;

Test #8: Mission→visions are mental models, not statements;

Test #9: Culture is built on shared, core mental models; and

Test #10: Learning constantly improves vision, mission, and culture.

If you follow tests #1-8 to develop your mission→vision, but you don't do the two implementation items (9 and 10) where *you build shared mental models into culture and encourage systems thinking and learning, the results will be lackluster.*

TEST #1: MISSION→VISIONS ARE SHORT AND SIMPLE

Perhaps the most important test of all is what I call the "9-year-old test." Your vision and mission should be simple, understandable, and easy (hence understandable to a 9-year-old). The fastest way to tell if an organization has an ineffective mission-vision or a good one is just how short it is. Most mission-visions are long, wind-bag paragraphs when they should be a few words. Most mission-visions wax on poetically in flowery prose when they should be simple. Bad mission-visions are paragraphs of platitudes; good mission→visions are short and sweet. Here's a great mission→vision (which happens to be our research lab's):

Mission: Engage, Educate, and Empower (E3).
Vision: 7 Billion Systems Thinkers (7BST).

E37BST

That string of symbols above (shorthand for our mission→vision) is a powerful organizational alignment tool. A leadership tool. A management tool. A culture-building tool. A make your organization adapt to a changing environment tool. E37BST is like a password or a slogan that's jam-packed with meaning to those who share in it. To us and the researchers who work for us, it's what gets us out of bed in the morning, tells us what to do that day, and keeps us on track. It's part rallying cry, part strategic direction, part instruction manual, part employee handbook. It is a mental model we all share and are all deeply motivated by and passionate about.

One of the most frequent questions we receive from organizational leaders is whether the mission→vision is for internal or external consumption. The answer, as you might have gleaned from the above paragraph, is that the mission→vision is first and foremost an internal document for transforming (or forming) an adaptive, effective, and powerful organization of systems thinkers. However an effective mission→vision will by definition be attractive to those outside your organization as well (including potential organizational members).

A well-crafted mission and vision can transform your organization and take it to the next level. But not if we don't change the way we think about things. See, we've been trained to distrust simplicity and adore complication. We think that a simple solution is too good to be true. Maybe it was all those snake-oil salesmen who were part of our Western heritage. Maybe it's that modern marketing is always promising us the silver bullet, knowing that it's part of our nature to hold out hope for one. Maybe it's because we've gotten burned. But no matter the reason, we've found that people will consistently take the complicated route over the simple one.

Time and time again, when we see great innovations, great inventors, or great teachers, they are the ones who focused their sight beyond the complicated to find the simple. Yet over and over, we see people settling for or believing that the only viable solution to their complicated problem is a complicated, over-engineered solution.

When it comes to your organization, this kind of thinking is dead wrong. If you can't explain your vision, your mission, and your culture in simple terms, then you don't understand it. And, if you don't understand your organization, then it will behave like an unruly teenager hellbent on making life difficult. Complicated is your enemy. Simple is your friend.

TEST #2: VISIONS CAPTURE A PICTURE OF A BINARY FUTURE STATE

Although it may seem redundant, visions are visual. This is something a lot of people miss. The word vision comes from the Latin *visio*, from *videre* "to see." The question is who or what is doing the seeing. A vision is something that a leader sees. Not because leaders see visions, but because those who envision can become leaders. We say *can* become leaders, because one additional thing is required: followers. At the precise moment when the first follower signs on, the person who had the vision transforms into a leader.

The basic idea then is that visionaries are people who see something that others may not see and then try to get others to see it. That might seem pretty basic, but the problem is we forget that this is the basic equation. We also forget that leadership is literally defined as someone who leads a person away from one place and toward another. Visions are visual pictures that show us the place we are going to: the proverbial promised land. *If* this is true, then it means that leadership can be highly distributed. If Frank sees a vision and becomes a leader when Sue decides to follow him and his vision, then it follows that Sue could help Tony see the same vision and Tony will follow Sue. Tony can then show it to Sally and so on. The basic act of leadership is to help people see something they didn't see and cause them to jettison their current activities and set off for a new goal. That's visionary leadership. It's not something that occurs in the beginning and then subsides. We need leadership whenever someone doesn't see what's possible and therefore isn't joining the movement toward it, or when someone who is part of the movement grows weary or loses sight of the vision. The leader is the one who goes back, puts their hand on a shoulder and points to the image, the icon, the logo, the visual thing that has a deep and abiding meaning to the weary ones and gets them moving again.

The visual nature of visions means that leaders need to spend some time talking about the picture of the future they see and how it differs from the current one. Differentiating these two pictures is every bit as important as differentiating one's products in the marketplace. It's a compare and contrast exercise: here's a picture of one place, here's one of another. Which one do you like? Yes, B. Well, I can show you how to get there. Thus, a visionary leader is born.

It's important to understand that visions exist in the future, not the present or the past. So combined with our understanding of how visions create leaders and followers and movements, we can see that the journey we are taking people on is from the present state (how things are today) to a future state (how they could be tomorrow). Visions are visual and they depict a future state so leaders must stand in the future and describe what they see.

Finally, a vision is a future state that is binary. It either exists or it does not. I'm either standing on the summit or I'm not. AIDS is either eradicated from Africa or it's not. We either are the market leader or we are not. This will become more important when we see that the third test for visions is that they are measurable, because binary is by definition measurable. For example, perhaps it is impossible to eradicate something completely, but the goal is to reduce it by 5%. That is binary because it either is or is not reduced 5%. "AIDS eradicated in Africa" is an effective vision. "AIDS decreased by 5% in Africa" is an effective vision. But "work on AIDS in Africa" is an ineffective vision because it's not binary (and not measurable).

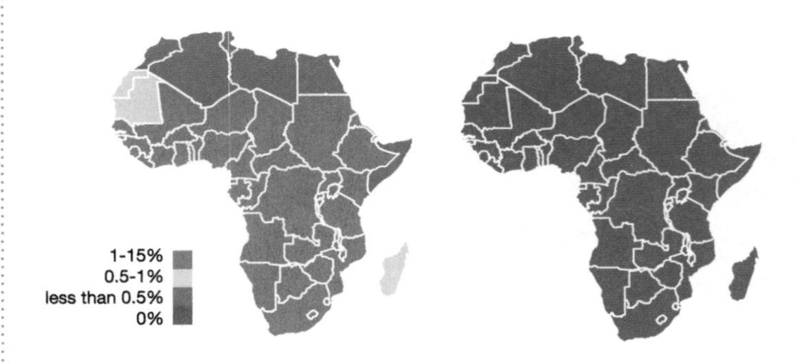

FIGURE 11.9: A Visual and Binary Vision

TEST #3: VISIONS ARE INTRINSICALLY MOTIVATING

Our organization's vision is our alarm clock and your vision should be the same. What we mean by that is that your vision should be so motivating it's the thing that wakes you up in the morning like a big dog licking your face. And on the mornings when you don't want to get up, it's the vision that motivates you to drag your sorry ass out of bed and get to work. That's the power of a great vision, it's intrinsically motivating.

There are countless ways to make extrinsic rewards work in your organization, but the one thing that must be intrinsically motivating is your vision. In sessions with teams where we help them work on their vision, people get goosebumps when they finally see their vision for the first time. People should be jazzed about it. If they're not, then your vision needs some work.

In the popular business book, *Man's Search for Meaning*, Victor Frankl explains that even things that we think of as supremely motivating, like happiness and pleasure, are less motivating than meaning.[7] Daniel Pink in his book *Drive* explains that a sense of autonomy and purpose far outweighs extrinsic incentives.[8] To be motivated, people crave autonomy, purpose, and meaning, and a succinct and powerful mission→vision delivers these things.

[7] Frankl, V. (2006). *Man's Search for Meaning*. Boston, Massachusetts: Beacon Press.

[8] Pink, D. (2011). *Drive: The Surprising Truth About What Motivates Us*. New York, New York: Riverhead Books.

This is why one of the best ways to find your vision is to ask yourself, "what pisses me off?" We always say that your vision is something that inspires. In many cases this is because it's the opposite of something that upsets you. So we often get clients started in the process by asking them about what pisses them off about how things are today. We start with people by drawing a line down the center of the board with the present picture of how things are and a future picture of how things are, as seen in Figure 11.10.

What do you see today?	What should we see tomorrow?

FIGURE 11.10: Finding Your Vision

Of course, this just gives you the seed of an idea for your vision, but the remaining tests will help you to refine it. Let's imagine a simple scenario to provide an example of how this process works. For example, imagine the result of applying Figure 11.10 is that today there are public beaches that have no lifeguards and it really makes us mad. We therefore envision a day where all public beaches have lifeguards. So we draw a picture of a beach without a lifeguard (an empty chair) and one with a lifeguard. That's our vision—that someday we could go to any public beach and see a lifeguard on duty. Vision: A Lifeguard on Every Beach.

The catalyst for this change will be your mission because it is the simple things that must be done repeatedly in order to achieve your vision. The next step is to connect the vision to a mission, which we call alignment or coupling. So what are the simple rules that must be repeated over and over again in order to get a lifeguard on every beach? This is just a logic problem. What needs to be done, over and over again, in order to bring about X? Hire, train, and deploy lifeguards. Do that over and over again and you'll have a lifeguard on every beach. Or you might argue that you will have to build public support in order to garner local resources (evangelize), establish standards and training for lifeguards (educate), and develop a global cause-based initiative that empowers beach-goers to organize locally to get a lifeguard to their beach (empower).

TEST #4: MISSIONS ARE SIMPLE RULES THAT FOLLOW A FORMULA

"We are what we repeatedly do. Excellence, then, is not an act, but a habit." - Aristotle

If a vision is the picture that we constantly see, then a mission is the thing that we repeatedly do. The mission of your organization is the simple rules of the complex system. Mission is the thing your employees (autonomous agents) do every day that will bring about the complex behavior you want that in turn will eventually achieve your vision.

We can't tell you how many times we see the ideas of vision and mission get butchered, conflated, and entangled by smart people. This is how you keep them separate. Visions are about seeing something: a simple, measurable, future picture of what could be. A mission is something we do repeatedly to bring about the vision, which is emergent. That is, if we do X,Y, and Z repeatedly, then our vision will emerge. So a mission→vision is just one big if-then statement. The X, Y, and Z part is the mission.

There is a simple formula that all mission statements must share. Even if all the elements are not always obvious in the final statement, they are there and this formula is critically important to know when devising a mission. It is like a MadLib. Every mission should:

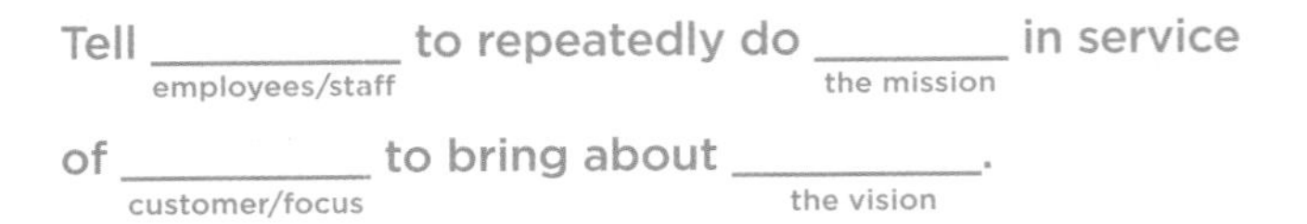

FIGURE 11.11: The Mission Formula

Built into this mission formula are some pretty remarkable ideas that may not be immediately obvious. First, alignment between the vision and mission is accomplished, which is the beginning of why VMCL brings about organizational alignment from top to bottom and interdepartmentally. The formula gives you a clear understanding of how the vision gets accomplished and how the mission and vision relate. The vision provides the focal point for the mission. The mission is the thing we do over and over again until the vision occurs. The vision is what we *see*. The mission is what we *do*. Pretty simple. The mission puts your vision on a pedestal, which is where it should be, because you need to get down to getting it (the mission) done.

The next thing that may not be immediately obvious is that the mission formula uses the simple word "repeatedly," because missions aren't something we do once. A mission is an algorithm for doing something over and over again. This notion is both related to CAS (simple rules) and the idea that your efficiencies (and profitability) will come from developing the capacity to do the same thing over and over again.

Finally, the mission formula is entirely customer-centric, which means your organization will be inherently customer-obsessed if you follow the VMCL path. It is often overlooked that the mission formula tells you what to do for whom and for what. The who is the customer and the what is your vision. So in actuality, the formula is balanced between doing what is mutually good for your customer and your organizational vision.

Because most of the mission→visions we work on with organizations are proprietary, let us use our own (Cabrera Research Lab's) to provide another example of how the formula is embedded in a simple mission statement.

> Our vision is: 7 Billion Systems Thinkers
> Our mission is: Engage. Educate. Empower.
> The people we serve include: individuals and organizations

But what our mission is doing implicitly shows that it follows the MadLib-style formula:

> "Tell [CRL employees] to repeatedly do [engage, educate, and empower] in service of [individuals and organizations] to bring about [7 billion systems thinkers].

Or the short version: E37BST

TEST #5: REPEATEDLY DOING YOUR MISSION SHOULD BRING ABOUT YOUR VISION

You can have a great vision and a great mission but they could suck together. This means that you need to pay some attention to how your vision and mission work together, or how they are coupled. Basically, the test for this is your answer to the following question:

If I ask 10 smart, rational, reasonable people whether doing my mission repeatedly would eventually lead to my vision, what would they say?

Visions and missions that are coupled work well together. They work in sync: the repeated doing of your mission should get you to your vision.

Here's a few examples of mission→visions that meet this criterion:

mission: Take one Step. Repeat.
vision: Standing on Summit

mission: Go Forth and Multiply
vision: Biodiversity

mission: Convert the Unconverted
vision: A [insert religion here] World

mission:[9] Get an intel chip inside every computer
vision: Intel Inside (every computer)

mission:[10] Engage, Educate, and Empower
vision: [insert number of students, staff, community members] Systems Thinkers.

missio:[11] Facilitate and Motivate Healthy Habits
vision: Living Healthy is the New Normal

mission:[12] Evangelize. Ensure. Enable
vision: The Power of Incentives Inside Every Company

mission:[13] Evangelize. Engage. Ensure. Energize.
vision: The World Subscribed

It should be noted that we have found, from working with hundreds of organizations, that similar organizations actually have very similar missions. For example, many product and service organizations need to sell people on the importance of their product/service (e.g., engage or evangelize in some way), get people using or trained in using their product or service (e.g., educate, embed, install, or incite the use of), and make sure the customer has independent, ongoing success with their product/service (e.g., empower). The more similar the organi-

[9] Intel Corporation's mission→vision

[10] Our mission→vision has been adapted and adopted by many school districts around the country using this basic format.

[11] MyFitnessPal Corporation's mission→vision

[12] Xactly Corporation's mission→vision. Note that although different organizations may use the same words in their missions, the meaning of those words may vary greatly. It is the meaning behind the words that matters.

[13] Zuora Corporation's mission→vision

zations, either in category or in purpose, the more similar will be their mission→visions. Most organizations have different visions, but by business type or sector, very similar missions. We have worked with many schools and school districts that use the very same mission (engage, educate, empower) and vision (X number of systems thinkers).

A mission is by definition the incremental, simple-rules path to achieving your vision. If following this path isn't getting you there, you either need to adjust the path or adjust the destination.

TEST #6: MISSION→VISIONS MUST BE MEASURABLE

Visions and missions are measurable. For visions, what this means is that you can determine whether you got there or not. The summit of a climb is a good example, because we can know whether we reached it or not. There's not a lot of room (maybe a few feet of undulating rock and ice) for debate. You either are standing on the summit and therefore the picture you envisioned in your head has now become a reality, or you're not. There are lots of different ways to measure something, including taking into account multiple metrics. Your vision might end up being slightly more complex than a summit but the point is that you devise some way to determine whether or not you arrived at your vision.

For missions, this means that each word or element in your mission can be counted in some way using some single metric or set of metrics. We advise that people keep these metrics fairly simple. You likely don't need a differential equation. Most of the time some simple counts will do. For example, for General Casey's mission we could choose to tie one or all of the following types of metrics to each part of the mission.

Table 11.4: Measurability of the Parts of a Mission

Parts of Casey's Mission	Examples of Metrics
Clear	# of: houses, blocks, streets, neighborhoods, towns cleared
Build	# of: relationships, partnerships, coalitions, hospitals, service projects built
Hold	# of: hours, days, weeks, months held

TEST #7: MISSION MOMENTS ARE RARE AND PRECIOUS

When we work with our clients, one of the ideas that they tell us is very powerful (and also initially a little hard to grasp) is the idea of "mission moments." The part of the idea that's somewhat difficult to grasp requires that you make a distinction between your mission statement/mental model and a single instance of your mission being fullfilled in real life.

Your mission is both a statement and a mental model shared by everyone in your organization. But because your mission is something you do in service of customers it is also contained within every individual interaction with a customer.

These instances where organization and customer interact are "mission moments." And here's the thing, everyone in your organization needs to cherish these mission moments. Maybe cherish isn't strong enough. Let's get a thesaurus.

Every person in your organization needs to see mission moments as something they: adore, hold dear, love, dote on, be devoted to, revere, esteem, admire, think the world of, hold in high esteem, care for, tend to, look after, protect, preserve, keep safe, treasure, prize, value highly, and hold dear.[14]

That's how you create a customer-centric company. Make mission moments reign supreme. Lionize them. Glorify them.

We do this because there's a very simple equation that governs your business regarding mission moments: if mission moments go well, you get more of them and your business thrives. If mission moments go poorly, you get less of them and your business dies.

TEST #8: MISSION→VISIONS ARE MENTAL MODELS, NOT STATEMENTS

When we help people create a mission→vision, one of the distinctions we make is between the vision *statement*, the vision *logo*, and the vision *mental model*, or alternatively, the mission *statement* and the mission *mental model.* Let's start with vision.

The vision *statement* is a string of words that describes the vision. In comparison, the vision *logo* is an iconic image that captures the vision. Neither is the vision itself. The actual vision is a *mental model*—the mental image that exists in the mind's eye of what the promised land looks like. Here's a simple example of our research lab's vision, which consists of a statement, logo, and mental model:

Vision *statement*: 7 Billion Systems Thinkers.

FIGURE 11.12: Vision Logo

[14] Lindberg, C. (2004). *The Oxford American Writer's Thesaurus*. Oxford: Oxford University Press.

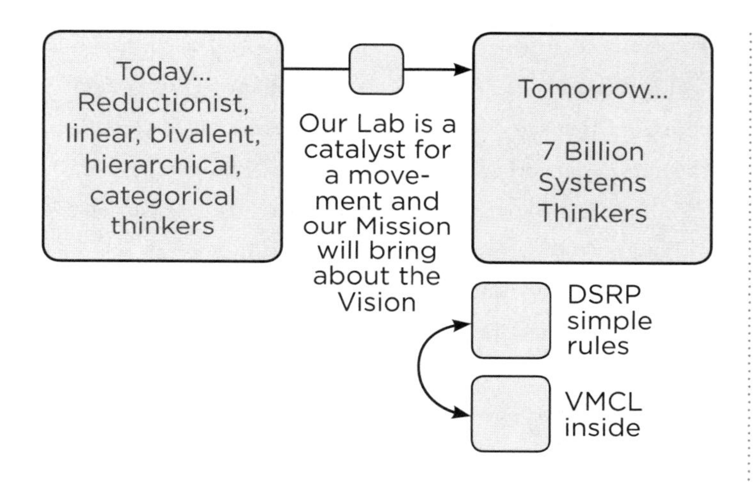

FIGURE 11.13: Vision Mental Model

This is so important and so often misunderstood. Lots of people equate their vision with the words on the page. Others believe that the words on the page have meanings derived from the dictionary. The *meaning* of the words on the page or the iconic logo of your vision is what the vision is. The shared mental model that every employee creates in their mind's eye is the vision.

For missions, we also need to make a few distinctions between: the mission *statement*, mission *logo*, mission *mental model*, and mission *moments*. For example, here's our research lab's mission elements:

Mission *statement*: Engage. Educate. Empower. (E3)

FIGURE 11.14: Mission Logo

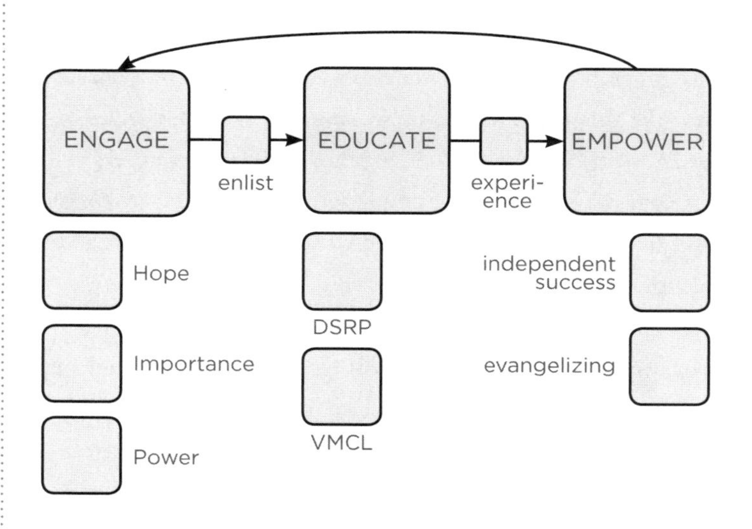

FIGURE 11.15: Mission Mental Model

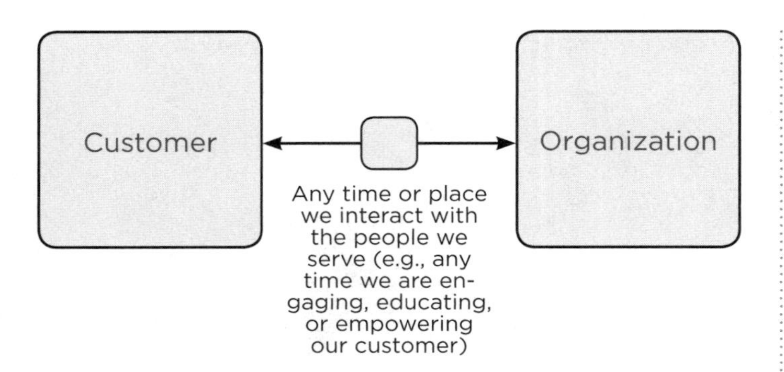

FIGURE 11.16: Mission Moment

The mission *statement* is just a few words on a page: Engage. Educate. Empower. A mission *logo* is simply a compressed, visual version of the statement and/or concept. The mission *mental model* is the concept that captures the *meaning* behind the mission. The mental model is the most important because it is the thing that people must *share* amongst themselves in order for the organization to actually *have* a mission. As you try to comprehend what our mission is, you'll likely go to the dictionary definitions of those words that you know and assume that we must mean that. But that's not the case. We don't use the dictionary definitions of those words, we use our own definitions; the words of the mission mean what we say they mean. Our mission *mental model* is the concept that people in our lab share—a shared understanding and agreement on what the three words mean (they mean *a lot* to us).

For example, *Engage* means that we focus on the "h.i.p." which refers to the hope, importance, and power of systems thinking. That's good marketing because it appeals to both the emotional and analytical aspects of the human psyche. *Educate* refers to the things we share—in business parlance, although we are a research lab, this is our product or service—DSRP and VMCL. *Empower* means that we focus on developing independent success in systems thinking and converting new systems thinkers into evangelists for systems thinking. This brings the cycle full circle back to *Engage* (the hope, importance, and power of systems thinking). Our mission concept has been so successful that it has spread beyond our own organization. That's a testament to the power of a finely tuned mission→vision.

Ten years ago, we were concerned that our Ivy League students could "do school" but lacked basic systems thinking skills like problem solving, creativity, emotional intelligence, and grit. We decided to take our research out of the tower and into the trenches. We started doing talks in schools around the country to anyone who would listen, and those talks really hit a nerve. Our concern about our students was shared among business, educational, and political leaders, as well as parents, teachers, and students. A year later, a principal named Christina Dickens from Fairfax County Public Schools in Virginia approached us with a novel request: "Can we, um, adopt your vision and mission?" Chris wanted to use our lab's mission→vision and culture-building strategies for her entire school, to train her teachers and transform her school into what she called a "systems thinking school." "Sure," we said, without realizing that this would lead to an E^3 thinking revolution in education. Today, districts, schools, and teachers across the USA are adopting our mission→vision as their own. Just as Chris's school is committed to engaging, educating, and empowering the systems thinkers the world needs, many other districts, principals, and individual teachers have taken responsibility for developing thinkers in their community with significant qualitative and quantitative results. Today, dozens of districts and tens of thousands of independent teachers have adopted our mission→vision. That's how powerful a simple mission→vision concept can be.

The mission moment is a relatively rare and precious moment—an event in time—when we get to do our mission. We cherish mission moments. You can tell that you are in a mission moment because it involves contact with your customer in some way. You might say to yourself that interacting with customers isn't rare because it happens every day. That is true, but compared to the many other daily activities your organization does, interaction with customers is a rarity. Consider, for example, all of the time a restaurant spends preparing for dinner service to the actual time restaurant staff engage directly with diners.

LEARNING→CULTURE

Making distinctions between your mission→vision *statements* and your mission→vision *mental models* is ridiculously important. It is important because it sets the stage for making sure that the mental model is shared. It's easy to share a mission→vision *statement*, just send it in an email to everyone in your organization. But that does not mean you have a mission→vision. You will only have a mission→vision when everyone in your organization *shares the same mental model* of your mission→vision.

Learning (L) and culture (C) are both predicated on the building and then sharing of mental models. So a discussion of mental models is critical.

TEST #9: CULTURE IS BUILT ON SHARED, CORE MENTAL MODELS

As leaders, we must both introduce people to the notion of mental models and differentiate them from reality. We need to teach that they are (often poor) approximations of reality. The next step is for us to see how often our mental models are out of alignment with reality. The simple rules of systems thinking (DSRP) help align mental models with reality.

The underlying problem is that people with deeply rooted biases often don't think of those biases as mental models that can be wrong or adapted. They don't even recognize that they have a mental model. They think that what they perceive is reality. So the solution is to think of mental models as being in between you and reality.

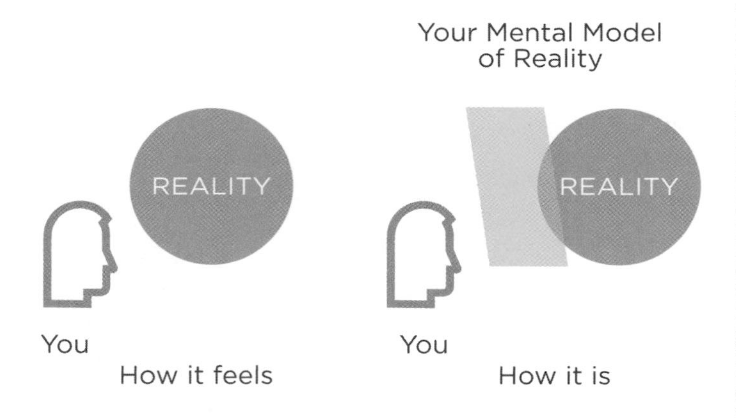

FIGURE 11.17: Acknowledging Mental Models

The mental models we have of our organization need to be more explicit. That's what understanding the traditional model of organizations is all about: making the model explicit so we can see and ask ourselves, "how often is this thing we rely on wrong?" So a true leader moves beyond articulation of a mission→vision by highlighting current difficulties and using the dissonance caused by the gap between the organization's vision and the current, difficult reality to inspire his or her team toward innovation.[15] Thus, a big part of creating your culture of learning is making people aware of the veil they wear every moment of every day. The veil of past experience, preconceived assumptions, and human biases. In summary, ensure that your employees understand mental models. Such an understanding entails 2 realizations:

1. Humans don't interact with reality directly; and
2. Humans indirectly relate to reality through our mental models of it.

YOU CAN GET INFORMATION BUT YOU HAVE TO BUILD KNOWLEDGE

Once you've laid the groundwork for the existence and importance and bias of mental models, you're ready to tackle another thinking error.. What we call an organization today is often nothing more than a bunch of people who have the same document, the same email address, share the same office building, and get paid out of the same bank account. That's not an organization. What makes a true organization is a group of people sharing the same mental models. Not all of the same models, but the important ones like vision, mission, culture, learning, and systems thinking.

Building and sharing the same mental models is no superficial task. It takes time, effort, repetition, and focus. But the alternative is that you have a bunch of people who share superficial attributes like where they work or what it says on their pay stub. That won't translate into a consistent customer

15 Senge, P. (2015). The Dawn of System Leadership. Stanford Social Innovations Review, Winter.

experience, a commitment to a larger cause, or a passion for work. To create an organization, we must do the hard work of getting people to share the same mental models. We need to identify what these mental models are, make them crystal clear and simple, and have a way to know whether people understand them.

Okay. So far, so good. Where the problem arises is when leaders think that covering information is the same as knowing it. If you have children, you know that teachers make the same mistake in school every day. They think that because they covered the material, the students know the material. Nothing could be further from the truth. Although we can transfer information, we can't transfer knowledge. People have to build or construct knowledge in order to possess it. Information is something you can get: you can get a phone number. But knowledge is something you have to build by structuring information to give it meaning. DSRP provides the universal modeling language to structure information.

This means that you need to spend more time helping your people to build shared mental models so that culture develops. Don't just hold a meeting and cover a bunch of new stuff. We'll soon talk about some of the most important mental models your people need to build (mission→vision).

In summary:

- What most of us call an organization is really a bunch of people who work in the same building;
- A true organization is a bunch of people who share the same mental models (i.e., culture);
- Shared mental models transform a group into an organization; and
- A shared mission→vision transforms your organization into a CAS or a superorganization.

Here's a classic example of the impact a powerful culture has on the results you seek.

Do you believe in miracles? In 1984 The US Olympic Hockey team beat better Soviet players to go on to beat Finland and win the Gold Medal. The win against USSR, called the Miracle on Ice, caused Americans to cry and go bonkers. Notice I didn't say that the US beat a better Soviet team, because they didn't. The USA was the better team. But by all accounts the individual players on the USSR side were pound-for-pound and stat-for-stat better. Most of the US players were collegiate and amateur athletes and the USSR had dominated in hockey both before and after the American win in 1980. And the same could be said for the other teams the 1984 USA team beat (Finland, Czechoslovakia, Norway, and Romania). What made the US so remarkable wasn't the individual players, but the team culture—a culture that started with the vision of

Coach Herb Brooks. The old saying that cash is king certainly has some merit. But culture is the real king. Culture is the wellspring of your competitive advantage.

FIGURE 11.18: Culture is King

But what is culture? If we say "apple pie," what do you think? You might have a grumbling in your stomach and think of a tasty treat. And that's what the rest of the world (those familiar with the treat) think, too. But if you're an American, then you think of more than a tasty treat. Apple pie is freedom. It's an idea: liberty and justice for all, Fourth of July, barbecues, family. Apple pie is a mental model that Americans share. It's part of our culture. Even if you've never met someone from the other side of the US, they have the same mental model as you do about apple pie. That's the power of culture. That's the power of shared mental models.

Whether it's something as simple as apple pie, or something as complex and hard to grasp as the USA's Miracle on Ice, culture is that somewhat intangible thing that binds people together. But contrary to popular myth, culture isn't some mysterious thing that only happens if we're lucky or only occurs over a long time. Culture is something that can be built. In fact, the idea of culture isn't intangible or mysterious at all, it is simple. Culture is what happens when people share the same mental models.

Culture is what happens when people share the same mental models. The vision and mission are not statements, they are mental models that need to be built by everyone in the organization.

If you want to build culture, build and share your mental models. Table 11.5 shows examples of mental models that might make up culture. For example, in the first column you have the core mental models that are shared across culture. In the second column are some types of mental models you might use that aren't core but will likely remain for a long time. For example, you might use *The Four Agreements*[16] as a cultural model to help people manage interpersonal relationships. You might also use a model such as CARE[17] or various other mental models built around rewards (President's Club) or mascots. Moving down in scale into everyday mental models that might change relatively regularly, your culture may share a model for meetings (Amazon and the US Military have banned PowerPoint), listening, or various other customs and norms of daily work.

Table 11.5: Examples of Different Levels of Shared Mental Models

Core Mental Models	Mental Models we hold dear and probably won't change	Everyday mental models that work but may change often
Mission→Vision (VMCL)	The Four Agreements	Meeting culture
Case for Existence[18]	CARE model	What we mean by listening
Systems Thinking (DSRP)	Iconic symbols or reifications of culture (e.g., mascot, awards)	Any customs or norms

The first and most important mental model you need to enculturate is the mission→vision. Remember, the vision and mission are not statements, they are mental models that need to be built by everyone in the organization.

[16] Ruiz, M. (1997). *The four agreements: A practical guide to personal freedom*. San Rafael, Calif.: Amber-Allen Pub.

[17] An example of a mid-level cultural mental model used by Xactly Corporation: Customer Focus, Accountability, Respect, and Excellence (CARE). Every employee at Xactly will be familiar with the four principles of CARE because, along with their mission→vision, it is a deeply held mental model of their organization.

[18] A Case for Existence is a powerful VMCL roadmap that is shared with every person in the organization. It summarizes the foundational principles, the before and after concepts used to develop the vision, the mission from internal (employee) and external (customer) perspectives, and the outcomes associated with the mission.

It's not a difficult idea. It's simple. But it will change you as a leader, change you as an owner, change your daily behavior, and change your organization dramatically.

The way we test whether CEOs are ready to hear what we have to say is we ask them a single question:

"Do you have a vision?"

The answer is almost always and emphatically, "Yes!"

"Oh, okay, what is it?"

The CEO says, "Well, um, it's, um, well it starts with how we are, um...well it says that..." (asks subordinate to get the vision)

Then we say, "Okay, that was a little rough. What would we hear if we asked the next person who comes down the hall what your vision is?"

The CEO says, "Well, I think you'd hear our vision."

We step out into the hall leaving the door open and wait for a passerby, "Excuse me, what is your company's vision?"

The passerby says, "Well, um, it's, um, well it starts with how we are, um...it's on our website. It's also on the wall in the waiting room, do you mind if I go get it?"

"No. That won't be necessary. Thank you."

CEO's don't like it when we do that. But they do get the point. They don't have a mission→vision. They have a bunch of words in a frame on the wall or on their webpage, which is not a mission→vision. They don't have something that is in the hearts and minds of every member of the organization. They lack this thing that motivates them when they're feeling lost and that clarifies where they need to focus every day.

Words mean nothing. Mental models, shared by everyone, create a culture committed to the mission→vision. A long paragraph of platitudes can't be understood and shared very easily and is even harder to implement. That's why short, sweet, and meaningful wins the day.

If every member of your organization cannot recite and explain your mission→vision, you don't have one.

Mission→visions aren't something you do because your B-school prof said you need them. It's something you do because your company can't be led and managed every single moment of every single day without it. A systems leader understands the importance of mental models to building culture. Therefore, the systems leader should remember the following:

1. Leadership means becoming a broken record: repeatedly clarify and help people build the most important mental models.
2. Leadership by framing: people often don't make the connection between the small, local picture and the big, global picture. Frame every act in terms of your VMCL.
3. Leadership is becoming the Chief Teaching Officer (CTO): learning is the engine of organizational effectiveness. Seize teachable moments.
4. Leadership by walking around (LBWA[19]): local action leads to the emergent properties you want, so get out and lead by walking around. When you do, be sure to be a broken record, frame things, and be CTO.

Notice that we didn't say that your most powerful weapon of all was your people. That's because they are not. People in and of themselves aren't that powerful, especially if they are bickering and nagging at each other. People who share a common mental model (i.e., culture) are powerful. People who come together under a common vision and execute a common and shared mission are powerful.

The only way to make sure that your mission→vision is aligned with your culture and in turn to totally align your organization is to enculturate the mission→vision into the hearts and minds of people. That means you will need to say it and show it 1000 times more than you thought was necessary.

Once you have your powerful mission→vision, then you have to lead. You have to ensure that the mission→vision doesn't remain a statement built of words, but rather becomes a mental model shared by every person in your organization. Mission→visions are not made up of words, they're made up of neurons or experiences manifested in mental models. They are ideas woven into the hearts and minds of your people. They are the fabric of your organization. They are your culture.

You need to create a culture campaign to enculturate your vision→mission. If you don't want or plan on doing this, then creating your mission→vision is a waste of your time. Because mission→visions are not words on a page—they are ideas in a mind, shared across a culture.

We have had a lot of success with our clients in creating mission→visions. The ten tests, a little patience, elbow grease, and group facilitation never fail us. But we'll be honest that some of our CEOs are so gleeful when they discover their new mission→vision that they are like children with birthday cake. They grab and go!

We don't think that's a bad thing because the CEO has been searching for something that succinctly captures what has

---

[19] First appeared as Management By Walking Around (MBWA) in Peters, T., & Waterman, R. (1982). *In search of excellence: Lessons from America's best-run companies* (p. 289). New York: Harper & Row.

been in his mind in a somewhat convoluted form for so long. He's so thrilled that it finally exists that he wants to get going, move on, and get to work with this new clarity. The problem is he has clarity, but no one else in the organization does. He spent the time building the mental model of the mission→vision in his mind and he sees what it means and how it's useful, but no one else does—yet.

It's that "yet" at the end of the last sentence that makes all the difference. When a CEO grabs his shiny new mission→vision and proceeds like Gollum in *Lord of the Rings*, we always feel a little sadness because we know that he won't quite get the value he paid for in his organization. But when a CEO grabs her mission→vision and sees it as a flag or coat of arms to be brought back to the organization and embedded as the centerpiece of the organization's culture, then we know everything is going to play out in a powerful way for her.

The mission→vision isn't done until it is woven into the cultural tapestry of an organization. The mission→vision isn't done until every person in the organization has had time to process it and build it into a mental model in their mind and their heart. The mission→vision is never truly done because through learning, our understanding of it evolves, new people come and go who need to build it for themselves, and so on. This process of enculturating the mission→vision is the epitome of leadership and can be fun and rewarding. To underline how important this step is we want to share a couple stories and then we'll show you four things you can do right away to start getting the mission→vision into your culture.

BUILDING CULT-LIKE CULTURES

Google, undoubtedly one of the most successful organizations to date, has coined the term "Goggleyness" to mean cultural fit. Interestingly, their CEO says that what they look for is people who are different because "diversity leads to great ideas."[20] He expounds by saying that cultural fit is more important than knowing how to do your job because if you fit the central tenets of their culture (humility and care for the environment of Google), you'll figure your job out over time.[21]

In his business classic, *Good to Great*, Jim Collins explains that organizations become great when they face the brutal realities of their business and build a focused culture.[22] He calls these cultures "cult-like" cultures because they have a rigidly enforced set of cultural norms (even if the norms themselves are flexible). In other words, those individuals who don't fit don't want to be there, and those people who do fit, love it.

As an organizational leader, it's your job to build this cult-like

---

[20] Bock, L. (2015). *Work Rules!: Insights from Inside Google That Will Transform How You Live and Lead.* New York, New York: Grand Central Publishing.

[21] ibid.

[22] Collins, J. (2001). *Good to Great: Why Some Companies Make the Leap...And Others Don't.* New York, New York: HarperBusiness.

culture by helping build the shared mental models of mission→vision and the other mental models that are important to it (see Table 11.5). Here's a set of simple rules for organizational leaders to follow:

1. Manage mission moments: create a system for yourself to expedite, ensure quality control, seize teachable moments,[23] and receive customer/employee feedback. Your focus and attention here communicate that mission moments are rare and precious.
2. Build shared mental models: capture mental models in MetaMap and LBWA these mental models. Use teachable moments and scrum[24] meetings to talk and build. Focus first, foremost, and forever on the mental models of mission→vision, culture, and learning by systems thinking. Next focus on your mid-level models including iconic models. Then focus on everyday mental models. Covering a mental model once is not enough! Frame every conversation and use every interaction as an opportunity to build culture. Your focus and attention here communicate that building a culture of learning together and developing systems thinkers is important.
3. Motivate and incentivize your team: allocate budget and develop an incentives list and give them to team members for pro-VMCL behavior and mission moments. Use these as party photos. Put your money where your mouth is.
4. Lead change effort: create a board at home with the change model on it. Create magnets of mental models and magnets of the people in your organization and move the people to the part of the MV graph where they live. Actively work on moving people into MV from not-MV.

Building a cult-like culture is important, and its cornerstone should be a laser focus on the mission→vision of the organization.

TEST #10: LEARNING CONSTANTLY IMPROVES VISION, MISSION, AND CULTURE

As an organization, you must be explicit about the purpose of the learning. Once again, the purpose of organizational learning is threefold:

1. To build your culture;
2. To make your mission go better, faster, and cheaper; and
3. To achieve your vision better, faster, and cheaper.

That's it. Now, as long as this focus is set, reset, and then stated explicitly, and then repeated, and then posted, internally marketed, and announced, you'll do great. Let employees search and find new and innovative ways to make your

---

[23] A teachable moment is a time when learning will be easiest and most relevant; it is usually *in situ* and experiential learning. The term was first used in Havighurst, R. (1953). *Human development and education*. New York: Longmans, Green.

[24] Adopted from the term scrum in Rugby football, a scrum in business usage is an agile, adaptive team process that uses short-cycle feedback to increase learning. The term first appeared in Takeuchi, H., & Nonaka, I. (1986). New New Product Development Game. *Harvard Business Review*, 86116, 137–146

mission→vision better, faster, and cheaper by looking in novel places for answers. But make sure the answer they find leads to a better, faster, or cheaper mission→vision.

If you are ever confounded about a problem in your organization, go immediately to a restaurant and pay attention. Restaurants are great analogs for business because unlike most business in which there is sometimes significant delay in feedback, there is very little delay in the feedback at restaurants. If a customer didn't like the food, they'll send it back or complain. If they don't like the service, they'll leave a lesser tip. If they didn't enjoy the meal, they won't be back and they'll tell their friends or post on Yelp. Restaurants are like little business simulators.

THE BEST CHEF DOESN'T COOK PARADOX

The best chef in a Michelin star restaurant often doesn't do any of the cooking. The best chef doesn't cook? That seems like a paradox.

If he's not cooking, what is he doing? He's standing at "the pass." In a professional kitchen, the pass is a special place designed for several purposes. First, it is the area where the head chef expedites, prioritizing and communicating orders as they come in. Second, the head chef uses the empty counter space of the pass to "plate"—the restaurant equivalent of quality control, ensuring that the fish isn't overcooked, the side dish is ample, and the final plating of the dish is aesthetically pleasing and clean just before it is passed to the server to be presented to the customer. Third, the head chef has a vantage point on plates as they are being bussed and returned so that he can see how clean they are. Are diners not finishing their meals? Bad sign. Is the food so delicious that every plate is coming back clean? Good sign. Fourth, the pass is a place for the executive chef to teach and for sous, meat, sides, and pastry chefs to learn. The executive chef knows that the success of his Michelin star restaurant rests not on his own ability to cook, but on his ability to get a number of chefs to meet his exacting standards and delicious taste profiles.

FIGURE 11.19: Systems Leaders lead From the Pass

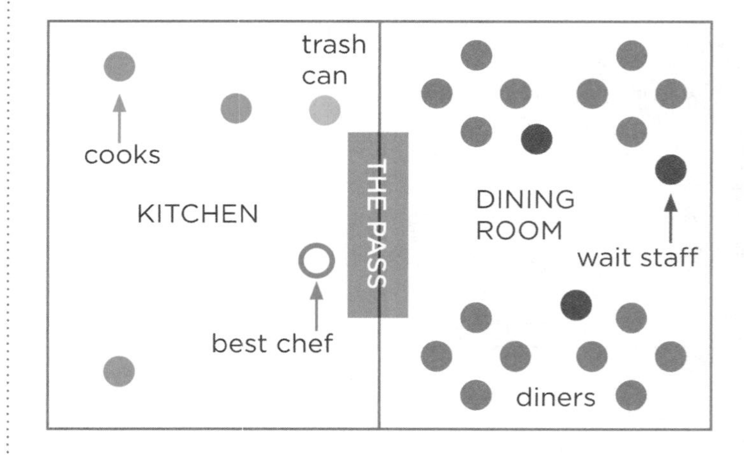

Organizational leaders fall into a trap similar to Michelin star chefs but often have not set up a sufficient pass to counteract it. They get promoted up the ranks for what they can do. Because they are competent. But at a certain point it's not about what you can do anymore, it's about what you are capable of getting others to do. Great leaders and managers make the transition from skilled doer to skilled teacher, because they know that learning ( ) is the organization's competitive advantage.

The goal of any CEO should be to create "distal CEOs"—people at the end of the myriad tentacle arms of an organization dealing with customers, suppliers, and partners, who are capable of meeting the exacting standards and connecting to the mission→vision held by the CEO. Organizational leaders need to do less and teach more. Focus on teaching people the mission→vision and culture, and get their leadership team to do the same. The return on your efforts will be substantial.

Figure out what the pass looks like in your organization (we've seen some cool ones!) and lead from it. It need not be a physical pass like in a restaurant but it does analogously need to have the same functionality: expediting and plating (quality control and service in mission moments), seizing the everyday teaching and learning moments, and getting real-world, ground-level feedback (how clean are their plates?).

Table 11.6 Summary Table of VMCL

Vision (V)	Visions are mental models that are: short, simple, binary, visual, a future goal, measurable, intrinsically motivating, the emergent property of missions, shared by culture, and improved by learning.
Mission (M)	Missions are mental models that are: short, measurable, repeatable, formulaic, simple rules that are shared by culture, improved by learning, and most importantly tell us what to do both to fulfill the vision and during rare and precious moments with customers.
Culture (C)	Culture emerges from the sharing of mental models large and small from mission-vision to cultural icons and mascots.
Learning (L)	Learning is a continuous process in which a CAS organization develops mental models and then tests them against reality. The feedback is then used to adjust the mental models.

RETURN ON INVESTMENTS

What's the return on investment (ROI) for implementing VMCL in your organization? VMCL:

- Simplifies: creates a single shared mental model of your organization that ensures everyone is on the same page;
- Focuses: ensures laser organizational focus on what matters most, especially when things get confusing;
- Aligns: links every organizational act and actor to the mission→vision (100k to 1k alignment);
- Becomes customer-centric: makes your organization totally customer-focused (happier customers=more revenue);
- Intrinsically and extrinsically motivates: motivates employees intrinsically where/when extrinsic/behavioral motivations fail (creating happier employees);
- Differentiates leaders vs. managers: clearly distinguishes leadership (vision/culture/learning) and management functions (mission);
- Maximizes efficiency: increases efficiencies (less expense) and guards against bureaucracy (less burnout/waste);
- Promotes adaptation: leverages human behavior and culture toward adaptation (big levers, big effect);
- Resolves organizational thinking errors: helps you avoid the big pitfalls of bad mental models;
- Applies across organization: can be used as a fractal or "project n" model (a universal, get-it-done-right tool); and
- Provides feedback: aligns mental models with reality in a constant feedback cycle. This is the key to adaptation and survival.

We must reiterate that VMCL doesn't implement itself. We've seen a lot of leaders waste a lot of time and money by religiously following tests #1-8 and then not paying attention to tests #9 and #10. Please don't make that mistake.

A learning organization is built by a set of individual systems thinkers who share a meaningful and common mental model of the organization's mission and vision. The leader must possess an ability to articulate the vision clearly, set forth a direct and connected mission, make the effort to build a culture entirely focused on the mission→vision, and most importantly, foster an ongoing reflective dialogue from which the organization can learn, adapt, survive, and ultimately thrive.

CHAPTER 12

CONCLUSION AND SYSTEMS THINKING MANIFESTO

CONCLUDING THOUGHTS

This book has discussed the importance of systems thinking—specifically the four simple rules of DSRP—in a host of realms, from science to daily life to leading organizations to developing ethical humans. In chapter 1, we saw how systems thinking is popular and promising because it offers people new hope for solving their wicked problems in a world that is increasingly complex. We introduced you to a statement that captures the essence of what systems thinking is all about and why it exists: *"wicked problems result from the mismatch between the way the real world works and the way we think it works."*

In chapter 2 we described the plethora of methods, concepts, theories, and approaches currently called "systems thinking" as the MFS Universe. We called all of this stuff Systems Thinking v1.0, and contrasted it to Systems Thinking v2.0, the subject of this book. We also discussed the significance of mental models, or the lenses through which we view the world. Most importantly, we introduced the idea that systems thinking is an emergent phenomenon that is ever learning and evolving through feedback from the real world. Systems thinking is a complex adaptive system (CAS) undergirded by simple rules.

In the third and most foundational chapter of the book, we discussed the four simple rules, DSRP, that underlie all forms of systems thinking. We noted how each rule involves the co-implication of two elements (for example, systems include parts, which imply a whole, and vice versa). DSRP patterns occur simultaneously rather than in isolation. For example, one can simultaneously identify the parts of a system, distinguish its parts, identify the relationships among the parts, and view parts from different perspectives.

Section two of the book was designed to help you become a better systems thinker, focusing in chapter 4 on visual techniques for conveying ideas, in particular mapping. We discussed the pros but mostly the cons of popular mapping techniques, and introduced MetaMap as a tool we think superior for both the conveyance of ideas and for making us better systems thinkers. That led us to chapter 5, which discussed cognitive jigs, structures that (like analogies) are repeatedly used to structure information and better convey meaning.

Chapter 6 focused on the predictive power of DSRP, which makes it easier to know where to look to discover new knowledge and insights. Chapter 7 explained that DSRP provides a more sophisticated, multivalent logic that society needs to solve contemporary wicked problems. Chapter 8 covered applying the simple rules to daily tasks such as making and reading tables and graphs all the way through understanding and improving existing methods of systems thinking and modeling such as system dynamics, soft systems methodology, and the increasingly popular science of network theory.

In section 3 we looked at how to scale systems thinking (chapter 9), a necessity if we are to harness its potential as an instrument of democratization. Then we used DSRP as a perspective on "designing" better humans and organizations (chapters 10 and 11, respectively). We saw that simple DSRP rules create not only highly intelligent systems thinkers, but also emotionally intelligent, prosocial people with an ethical compass. We explained how each rule aligns with and reinforces a principled ethical stance. With respect to organizations (chapter 11), we explicated how systems thinkers can design better organizations (CAS) by focusing on attracting and incentivizing the best agents and inculcating them in a culture based on simple rules of the mission that (followed daily) achieve the organization's vision. The rules are Vision, Mission, Culture, and Learning (VMCL).

So let us return to our question of "why systems thinking?" Systems thinking is seeing the underlying structure of systems and thought. Systems thinking isn't about content—it is content agnostic. Thus it can be applied to *any* content or topical domain. This broad relevance combined with the simplicity of DSRP rules makes systems thinking inherently democratic. That is of primary importance.

Secondly, in an increasingly complex, information-inundated, fast-moving world, our problems are increasingly wicked, making them seem intractable. In a different context, Einstein argued that we will not succeed in solving our current problems using the thinking associated with their origin.[1] It was indeed our thinking that created today's wicked problems. They result from the mismatch between how real-world systems actually work and how we think they work. DSRP helps us by revealing the structural pitfalls in our mental models.

The promise DSRP holds for solving issues from everyday life to socio-economic and political morasses that threaten our society and planet cannot be ignored. For those seeking answers in systems thinking, far too many are being turned away from its immense promise by a deluge of confusing, needlessly complicated answers. This is both immensely unfortunate and entirely unnecessary, because the underlying DSRP rules are simple.

[1] (1946, May 25). Atomic Education Urged by Einstein. *New York Times*.

Yet these simple rules capture complexity in a way that aligns with the reality of our contemporary world and its wicked problems.

It is perhaps ironic that the simple rules we prescribe—DSRP—are intrinsic to our thinking. It is innately human to make distinctions, to group things together, to take perspectives, etc. *What is not innate is to execute the simple rules consciously and to their full extent.* For example, we need to not just adopt a perspective, we must be both conscious of doing so and of omitting other perspectives.

An important yet perhaps underemphasized aspect of systems thinking v2.0 (DSRP) is its inherently democratic nature. First of all, contrary to what an examination of systems thinking v1.0 might yield, v2.0 is not just for the library or the lab. We don't need more scientists, we need more scientific thinking among the populace. We have already discussed how inherently ethical and humanistic it is to acknowledge the other created by our distinctions and to take the other's perspective. Similarly, our humanity is distinguished by a need to belong to something greater than ourselves, to recognize that we are parts of a larger whole, and to appreciate the nuances of our intricately interconnected relationships.

Within different sectors of society, be it academia, religion, politics, etc., immersion can breed insularity. Some religions, creeds, political groups, and even some disciplinary silos also elevate certain humans over others. Such problems are incongruous with DSRP, which abhors violent distinctions ("us vs. them"), encourages multi-perspectival thinking, and encourages appreciation of our interrelatedness. A code of ethics and a guide to living inhere in the simple rules. The more you practice DSRP, the less susceptible you are to manipulation, and the more likely you are to realize your own human potential.

At the same time, DSRP's application transcends the individual level, guiding us to create CAS super-organizations of systems thinkers. How can one lead a system? By building a culture that focuses on the simple rules that system thinkers must follow to achieve your vision. This advice applies from the corporation to the small business to the PTA. Unlike many organizational guides and maxims, VMCL has traction, is simple to implement, and lets everyone know what to do every day.

Similarly, what gives systems thinking teeth are the simple rules. From our personal lives to politics, the outcomes we seek are invariably complex and not susceptible to direct control (or manipulation). To achieve our desired ends, we must focus on the process. Simple rules repeated often. Simple things done by many lead to big changes. DSRP is an action agenda for science, for life, for everyone, just for today and for every day. It can help us be—both individually and collectively—more adaptive, better learners, clearer thinkers, and better humans. DSRP is the unified basis for human cognition and holds the common solution for the wicked problems that imperil our society.

DSRP is systems thinking made simple. That simplicity is beautiful, elegant, sophisticated, and powerful. From four simple rules done over and over again emerge marvelous, adaptive solutions and outcomes.

If you enjoyed this book, please share it with others and join us in our mission to Engage, Educate, and Empower 7 Billion Systems Thinkers. We leave you with our manifesto. ❋

ALL PROBLEMS RESULT FROM THE MISMATCH BETWEEN HOW REAL-WORLD SYSTEMS WORK AND HOW WE THINK THEY WORK.

THINK BETTER SOLUTIONS, SCIENCE, FAMILIES, SCHOOLS, BUSINESSES, GOVERNMENT, SOCIETIES.

BECOME A SYSTEMS THINKER

FOLLOW 4 SIMPLE RULES.

MAKE DISTINCTIONS AND RECOGNIZE SYSTEMS, RELATIONSHIPS, AND PERSPECTIVES (DSRP).

MIX AND MATCH THESE RULES LIKE PRIMARY COLORS.

SYSTEMS THINKERS CHALLENGE BOUNDARIES, SEE INTERCONNECTIONS, AND ARE PART OF A LARGER WHOLE.

WHEN YOU CHANGE THE WAY YOU LOOK AT THINGS, THE THINGS YOU LOOK AT CHANGE.

SYSTEMS THINKING IS A NEW ETHOS.

SMALL THINGS DONE BY MANY CAN LEAD TO BIG CHANGES.

WHEN WE TAKE THE TIME TO THINK ABOUT THE WAYS WE THINK, IDEAS THAT CAN CHANGE THE WORLD BECOME POSSIBLE.

Videos (Copyright)

Tables

Figures[1] (Copyright)

[1] All figures courtesy of authors unless otherwise credited.

Cover images: Shutterstock and the authors